LIGO: Prelude to Revolution

Conversations at Tahoe

LIGO: Prelude to Revolution

Conversations at Tahoe

by

Ed Hatch

Copyright © 1999 by Edwin E. Hatch

All rights reserved. No part of this book may be reproduced, stored in a retrieval system, or transmitted by any means, electronic, mechanical, photocopying, recording, or otherwise, without written permission from the author.

ISBN 1-58500-794-3

1stBooks - rev. 01/21/00

About the Book

The two Laser Interferometer Gravitational-wave Observatories (LIGO) are designed to detect gravitational waves and are expected to provide a powerful new tool for exploring the universe. They will not. Clearly, gravitational waves are real — but the LIGO observatories will not be able to detect them. This book tells us how and why this LIGO failure may ultimately trigger the most radical change in theoretical physics since Einstein's introduction of special relativity in 1905.

Foreword

In 1989 I first encountered the work that has since become my brother Ron's, Modified Lorentz Ether Theory (MLET). It intrigued me so much that I haven't really been able to get away from it since.

Among other things, the theory makes some definite predictions with regard to the current efforts to detect gravitational waves. The United States alone has spent well over two hundred million dollars on the two Laser Interferometer Gravitational wave Observatories (LIGO). Many other countries have launched their own efforts. It's a really big deal. If you're not aware of just how big, or would like to know more about the gravitational wave detection efforts, go to the LIGO home page at www.ligo.caltech.edu.

The author of the Modified Lorentz Ether Theory is my brother, Ron Hatch. At this writing (Summer 1999) Ron is the chair (an elected office) of the Satellite Division of the Institute of Navigation. This is the organization that conducts the premier conference relating to GPS, a conference typically drawing over 2,000 people to the sessions and exhibits.

Ron has been working with navigation and communications using satellites since 1962, when, still in college, he worked for the U.S. Science Exhibit at the Seattle World's Fair demonstrating the Doppler effect on the signals received from the TRANSIT satellites of the Navy Navigation Satellite System. This system was developed by Johns Hopkins Applied Physics Laboratory, where Ron worked developing navigation algorithms immediately following college.

In late 1993, after many years of working for others, Ron began what proved to be a very successful private consultation practice that included such clients as NASA, FAA, Motorola, and Leica, the Swiss survey company. In 1995, he, along with four other consultants, started NavCom Technology, which has grown into a successful GPS and satellite communications company employing more than 50 people. Although full time at NavCom since 1997, Ron has maintained some independent consultation work with others.

In 1994, Ron was awarded the Johannes Kepler award for 'Sustained and Significant Contributions to Satellite Navigation' from the Satellite Division of the Institute of Navigation — only the fourth recipient of this prestigious award. These 'sustained and significant' contributions were made over many years. He's received eight patents relating to GPS and satellite navigation, with several more in process. Among contributions not patented, is a technique Ron developed for removing much of the noise caused by electromagnetic reflections from the fundamental Global Positioning System measurements. This technique is now employed in virtually all GPS receivers and is referred to in FAA algorithm documents as the 'Hatch filter.'

Ron's understanding of the GPS system includes awareness of the effects of gravity and velocity on precision atomic clocks and other important relativity effects. His Modified Lorentz Ether Theory came from a driving preference for rigorous understanding, rather than from any personal desire to significantly impact theoretical physics. But in the course of that effort he became convinced that acceptable understanding of relativity effects could only come with a radical departure from consensus thinking. MLET, in its current form (including the prediction that LIGO will not detect gravitational waves) has far surpassed his original, rather modest goal.

This book is, primarily, an attempt to make MLET's prediction with regard to the gravitational wave observatories known before they are fully operational. Secondarily, it is an attempt to introduce the lay public to the theory itself.

This introductory attempt is structured as a series of conversations that never really took place. While the story is fictional, the theory is real, the predictions clear.

Enjoy it. Think about it.

Ed Hatch
2 August 1999

Gravitational Waves
[Steven Mitchell and David Rhodes]

10 May, 1999

→ Let's start with an observation Einstein[1] made in a speech in 1920:
> It would be a great advantage if we could succeed in comprehending the gravitational field and the electromagnetic field together as one unified confirmation.

The LIGO observatories will soon force a reconsideration of the factors that have, to this point, made this crucial unification impossible; and consequent serious re-examination of basic assumptions should trigger what will ultimately become the biggest revolution in consensus thinking since 1905.

♣ How can observing gravitational waves do that?

→ Observing gravitational waves can't — but the observatories will. Years of null results should eventually lead to serious questioning of the current consensus. We now know that the observatories will never 'observe' a thing.

♣ Not detect gravitational waves? That's crazy!

→ They'll find nothing.

♣ So we're throwing away hundreds of millions of dollars? Get real. The LIGO observatories will do what they're designed to do.

→ They'll detect nothing at all. What they're —

♣ The magnitude of the effort is a good indicator of the level of confidence. Look: Americans have the LIGO observatories and have proposed a space-based LISA; The French and Italians have VIRGO; The Germans and the British have GEO 600; The Japanese, TAMA 300; Australians, ACIGA; Brazilians, Graviton. All specifically in support of the observation of gravitational waves.

→ Impressive, isn't it? A lot of people will be disappointed. What they're looking —

♣ Impressive? They know what they're doing. The results aren't in doubt.

→ Results aren't determined by consensus. They'll detect nothing, because what they're looking for —

♣ You think the detectors aren't sensitive enough?

→ Sensitivity isn't the problem. What they're looking for just doesn't exist.

♣ Your ignorance is showing. Both theory and observation demands them. How else do you explain the energy loss in the observed orbital decay of the Hulse-Taylor binary pulsar?

→ Oh, gravitational waves exist — no one doubts that. Orbital energy is being lost — clearly radiated away.

♣ First they don't, now they do. I agreed to listen, but this is ridiculous.

→ I never said gravitational waves don't exist. I said what they're looking for doesn't exist.

♣ And they're looking for gravitational waves.

→ They're looking for a certain type of wave, a type of wave that doesn't exist. Current theory makes three assumptions regarding gravitational waves: (1) Gravitational waves exist. (2) They exist in a form popularly described as 'ripples in space-time'. (3) They are energetic enough to be detected.

- ♣ Good assumptions.
- → Current theory says they are. But our Modified Lorentz Ether Theory —
- ♣ Ether Theory? An ether theory?
- → Sorry. Let's save that discussion for later. Your objection to an ether is quite easily —
- ♣ Do you really expect me — or for that matter any knowledgeable physicist — to take you seriously? An ether. You're paying me thirty thousand dollars to listen to this kind of nonsense? Somebody's intent on throwing away good money.
- → The ether is essential to good theory. You're right in one respect though — as George Gamow[2] put it:

  > Albert Einstein became the ruler of modern physics by cutting the ethereal knot with the sharpness of his logic, and throwing the twisted pieces of world ether out of the window of the temple of physical science.

  So, no, as you say, we shouldn't, and don't, expect that serious physicists might waste their time on such nonsense. Why do you think we're paying you to listen?
- ♣ I'm not sure you're paying me enough for this!
- → The ether's real — much more real than your own non-vacuous vacuum. But let's get back to the gravitational wave discussion.

  I pointed out that current theory makes three assumptions regarding gravitational waves: (1) They exist. (2) They exist in a form popularly described as ripples in space-time. (3) They are energetic enough to be detected. You agreed.
- ♣ I did. The assumptions are based on well understood phenomena.

→ MLET (our Modified Lorentz Ether Theory) tells us that the first assumption is valid, gravitational waves clearly exist. And there's no good reason to doubt the third assumption, the radiation surely can be quite energetic.

♣ So, why won't they find what they're looking for?

→ The problem is with the second assumption. MLET, in providing Einstein's long sought 'unified confirmation' of the 'gravitational field and the electromagnetic field,' tells us that the gravitational waves are not of the expected form.

♣ Oh, great. A brand new wave form. A Nobel Prize for someone, I'm sure.

→ It's current theory that requires the extraneous wave form. The very real gravitational waves are identical to electromagnetic waves.

♣ That's preposterous.

→ Why two when one will do? Nature tends to be conservative, don't you think?

Seriously, MLET very clearly tells us that these gravitational waves are, in fact, identical in every sense to the waves that we commonly refer to as electromagnetic waves — interacting compressive (gravitational) and shear (kinetic) distortions of the ether.

♣ And you believe that?

→ I believe that. Just look at *why* current theory is forced to consider them two different types of waves — why Einstein was forced to give us divergent explanations of the gravitational field and the electromagnetic field in the first place. Einstein[3] wrote:

> Since the special theory of relativity revealed the physical equivalence of all inertial systems [his

absolute symmetry at all levels] it proved the untenability of the hypothesis of an ether at rest. It was therefore necessary to renounce the idea that the electromagnetic field is to be regarded as a state of a material carrier. The field thus becomes an irreducible element of physical description . . .

Without some medium supporting electromagnetic waves, Einstein was clearly forced to make the fields, in themselves, irreducible. But gravitational waves propagate through a medium as distortions of that medium (ripples of space-time curvature). Irreducible fields propagated through a vacuum in the one case, ripples in space-time in the other — very different waves, but propagated at identical speeds.

On the other hand, with a medium supporting electromagnetic waves, the electromagnetic fields are no longer irreducible but are seen as elastic distortions of that medium. Numerous considerations make it very clear that gravitational waves are composed of the same kinds of distortions — both are gravitokinetic waves, compressive and shear distortions, in phase, both propagated at the speed of light. In other words, gravitational and electromagnetic waves are the same kind of 'ripples in the ether'.

♣ So you simply discard special relativity? Not smart. Paul Davies[4] has reminded us that,
> If relativity were wrong, our detailed understanding of subatomic physics would collapse. . . From quarks to quasars, scientists would no longer be able to understand the basis of their own immense knowledge.

→ Davies is wrong. With the transition to MLET scientists are now "able to understand the basis of their own immense knowledge," for the first time ever in very many particulars. Since current theory is, in a sense, little more than the skeleton of MLET — MLET arbitrarily stripped of physical meaning — MLET's restoration of a physical model is in no way destructive of understanding.

♣ How could even an idiot make such a ridiculous claim?

→ Current theory can't tell us much beyond the raw experimental data and experimentally verified mathematical relationships. Stephen Hawking[5] expressed this impotence rather succinctly when he wrote:

> I take the ... viewpoint that a physical theory is just a mathematical model and that it is meaningless to ask whether it corresponds to reality.

MLET, on the other hand, gives us a consistent physical model that allows the derivation of the current mathematics. This was Lorentz's position when, prior to Einstein, he derived the mathematics (the Lorentz equations) supposed to support special relativity. And the reason Einstein rejected Lorentz's physical model is clear — Einstein[6], in a speech in 1920, stated:

> For the theoretician such an asymmetry in the theoretical structure, with no corresponding asymmetry in the system of experience is intolerable.

Einstein's positivism is showing here. It wasn't the mathematics or the experimental data that dictated the rejection of Lorentz's sensible, visualizable physical reality — it was nothing more than a quaint personal notion that a reality that in any way exceeds direct experience is intolerable.

Virtually the whole difference between MLET's visualizable, sensible reality and current theory, arises from Einstein's rejection of Lorentz's underlying asymmetry.

- ♣ And, I suppose, you'll show me how this sensible reality is reflected in the current mathematics and is consistent with our massive experimental base? A pretty big order, if you ask me.
- → Granted — it is a large order. That's why you're here for a month.

    Again, the point I really want to emphasize today, is that the sensible reality clearly tells us that the LIGO observatories will never detect gravitational waves.
- ♣ You expect me to believe this nonsense?
- → Me? No. I understand where you're coming from. I'm personally fairly sure you won't be able to question SRT, no matter the strength of the evidence. But I do hope that a few years from now, when LIGO scientists are struggling with the frustration of not being able to detect what they know is out there, you'll remember that there's a very good explanation for the null results.
- ♣ You're different than most nut-cases — I'll grant that. As crazy as they come, but different. It's very unusual for people like you to stake all hope of credibility on an unambiguous prediction that'll soon be tested.
- → Good theory should do that. Don't you think?
- ♣ Good theory? Yeah, sure.
- → Stay with us and we'll show you just how good it really is.
- ♣ You really depress me.
- → What's to be depressed about?
- ♣ You're willing to spend millions of dollars to argue that the physics community doesn't know what it's talking about and you wonder why that depress me?
- → Science is about change. Significant challenge to consensus shouldn't depresses you.

♣ Surely even you can understand the danger to good science. I'm not the best known theoretical physicist by any means, but based only on my limited credibility, I must get at least two or three proposed papers a month from crackpots like you with some new theory that shows how ignorant we mainstream physicists are.

→ And you simply toss them — as any good physicist would? Right?

♣ I do, of course. But I think Leon Lederman[7] best expressed the real concern with regard to this glut of nonsense when he wrote:

> The tragedy in all this is not the sloppy pseudoscience writers, not the Wichita insurance salesman who knows exactly where Einstein went wrong and publishes his own book on it. It is the damage done to the gullible and science-illiterate general public, which can so easily be duped.

You'll get the treatment you deserve from the physics community. But the general public can't so easily be protected from well-funded kooks — and if you compromise the general public's trust in informed consensus, then support for important experiments may be jeopardized.

→ You shouldn't worry, there's a solid majority that isn't all that gullible.

♣ Do you think the gravitational-wave experiments would've been funded at the current level if you'd come forward with multi-million-dollar backing a decade or two ago claiming, as you do now, that they are worthless?

→ Worthless? Well, the author of MLET did once say that he believed the huge amounts of money being spent on gravity-wave detectors were being wasted. But that was said with respect to the stated aim of the experiments — to eventually use gravitational waves to detect things that couldn't be detected in other ways. Paul Davies[8] wrote:

> Using gravity-wave detectors as 'gravity-telescopes' is on the horizon. With such a facility we could 'see' into the dense hearts of quasars and neutron stars, probe to the very edges of black holes and maybe eventually listen to the rumble of the primordial big bang itself.

If one sees this as the only benefit of the observatories, then sure, they're worthless. But from our perspective they're far from worthless.

♣ What do you mean? You claim to know they'll never detect gravitational waves.

→ And you claim to know they will. Who's right? The situation demands experimental resolution.

♣ Very funny. There's no 'situation' to resolve. The observatories were designed to observe — they're 'observatories' not experiments designed to test your ridiculous hypothesis.

→ But they will. And that makes them important.

♣ Their importance has nothing to do with your absurd theory.

→ It's a shame that it takes hundreds of millions of dollars to prove something that simple re-examination of basic assumptions could probably accomplish just as well. But given the current environment it's probably money well spent.

♣ I'm curious — what can you hope to gain from my listening to this nonsense?

→ Surprising as it might seem to you, we do have credibility problems — so we're paying to be heard. Our situation reminds me of something else Davies[9] wrote (in the same *Why Pick on Einstein?* article you quoted from earlier, if I'm not mistaken):

> Most editors of science magazines and journals make special provision for coping with the huge influx of papers and letters, many bearing private addresses in California, purporting to disprove or improve Albert Einstein's monumental work on the theory of relativity.

You may have noticed that the address currently used by our Tahoe Project is Jim Price's vacation home address — a 'private address in California.'

♣ But we're in Nevada, aren't we?

→ Yeah, we're in Nevada now. But those of us doing the initial planning worked from Price's vacation home near the base of Squaw. Happily for us, Price developed this condominium resort on the Nevada side of the lake. We took over this building a few days ago, so we'll soon be free of the terrible stigma of a "private address in California." Unfortunately, that *still* leaves a slight credibility gap.

♣ Unfortunately, fifty million dollars can keep even idiots going for quite a while.

→ The money helps. But let's get back to our credibility concerns and why you're here.

Seriously, have you ever wondered why there's so much noise, so much nonsense floating around out there?

♣ It's rather obvious — a lot of ignorance coupled with inability to accept the non-intuitive implications of Einstein's special relativity theory — too many people like you.

→ There's a lot of ignorance, certainly. But, beyond that, current theory is extremely vulnerable — it's by no means the best we can do. Intelligent people sense that. Modern physics is a mess.

♣ And by the end of my month here, I'm supposed to see the light — reject this 'mess' in favor of some new physics?

→ That'd be nice. Unrealistic, but nice. We're interested in whether, or to what extent, we might be able to create in you an ability to question the current consensus.

♣ You've certainly made one point — you're certifiably insane. Tell me something. In some respects you seem reasonably intelligent — so what in the world convinced you that you understand physics better than Einstein or, for that matter, any modern physicist.

→ We've never made such a claim and you know it.

Personally, my education and experience has been in the computer sciences. I'm no physicist, but my experience has helped convince me that MLET is essentially correct.

♣ And how, pray tell, might your experience be relevant?

→ Some types of experience are applicable across disciplines.

♣ Brilliant. And expert knowledge and expert opinion are only incidentals?

→ They're always essential of course. But in today's world, facts and informed opinion are readily available to anyone.

In trouble-shooting many problems over a number of years, I learned something I think significant. When very knowledgeable and very intelligent people are faced with seemingly intractable problems, there's a very good chance that those problems arise from a single, simple, unwarranted assumption in an area supposedly known to be error free.

♣ And your point is?

→ In such a situation an outsider has an important advantage — the outsider can (must, really) question known truth. This advantage is real.

The more intractable the problem, the less likely it is that the insider, the resident expert, will find the time or have any inclination, to question that which is personally known to be true, to be error free. Very rarely will one re-examine fundamental assumptions once they are no longer seen to be assumptions — once they have become settled truths.

♣ So, because you're not a physicist, you've been able to find our silly little error and solve all our problems?

→ I'm not the author of MLET. But the fact that the author was not a physicist probably did have much to do with his asking the right questions. And the result does solve most of the really difficult problems. So yes, by simply discarding one unwarranted assumption from current theory's unquestioned truth, significant error was eliminated and problems were solved. And, since every physics student is solidly grounded in the firm conviction that there can be no problem in precisely the area where the error resides, it's highly unlikely that any professional physicist would've asked the right questions.

♣ So we needed an outsider?

→ At the least, you needed someone who could allow himself to question the unquestionable.

♣ And now that the error has been found?

→ The physics community can probably handle it from here.

- ♣ This is a silly conversation. You really think you've given physics something that no physicist could, don't you?
- → MLET required questioning current theory in the one area where questions are clearly not tolerated within the consensus community.

    And the questions yielded answers. MLET replaces both special and general relativity and confirms the solid foundation of quantum theory — with remarkably little change to the core mathematics in any area. And in doing this, MLET provides a degree of universal convergence that is clearly impossible within the constraints of current theory.
- ♣ Sounds like we have a Grand Unified Theory and Theory of Everything rolled into one.
- → Closer to exactly that than anything you've ever seen before.
- ♣ And the brilliant author of this all-encompassing theory has developed every one of the very many new interpretations required by the sweeping scope of the simple change?
- → Actually, as the 'Lorentz Ether Theory' portion of the name suggests, a very large part of the theory has been around for decades in bits and pieces — pieces largely ignored because they generally do conflict with SRT. However, the 'Modified' does supply some key concepts, concepts that provide the essential unification, the universal convergence. And so, in the Modified Lorentz Ether Theory, we now have a fully convergent, universal theory, that provides (among many other things) the following specific improvements:
    - A superior explanation of the null results of the

Michelson-Morley experiment.
- A superior interpretation of the nature of electric, magnetic and gravity fields.
- A superior interpretation of the nature of mass and of inertia.
- A superior description of gravitational (electromagnetic) waves.
- A superior interpretation of the nature of matter.
- A superior interpretation of the effects of velocity on measurement and on matter.
- A superior interpretation of the effects of gravity on measurement and on matter.
- A superior interpretation of the speed of light as a limit.
- A superior interpretation of the nature of the vacuum (the ether).
- A very superior explanation of the Sagnac effect (optical gyroscopes).
- A superior interpretation of the so-called twin paradox.
- And all of this with no disparate domains, no 'special' theories.

♣ How modest can you get? Superior in every way. But you must admit that current theory works quite well.

→ Yes it does, and from the MLET perspective we can see *why* it does. Actually, given nearly a hundred years of very good experiments, it would be exceptionally naive to think that theory would be allowed to significantly deviate from experimental necessity. So yes, current theory works. With the aid of very good experimental data we've established valid relationships and so derived rather good mathematical models. It's not surprising that, pragmatically, it makes little difference whether one stays with the old or switches to the new — the experimental data is necessarily identical, the mathematics nearly so.

Look at all the areas where I claim MLET superiority. In very few cases does the superiority arise from improved ability to predict direct measurements.

♣ So, only metaphysical differences? Why are we wasting our time?

→ Metaphysical? Look at what you're calling 'only metaphysical.'

Theory must, first of all, work. But good theory should do much more than that. Current theory gives us a good mathematical structure but the interpretation defies any attempt to comprehend any independent reality. The new theory uses almost identical mathematics but gives us an intuitive understanding. On the one hand we can't even talk about the meaning of reality apart from the mathematics, while on the other we have an intuitively visualizable reality from which we can *derive* the mathematics.

Both work, but only one provides understanding. Given just equal ability to predict and equal consistency with experimental data, which should we prefer?

♣ You're claiming an awful lot.

→ Clearly. But, as a matter of fact, MLET doesn't just work as well as current theory, but does, as I've claimed, provide superior ability to predict and superior consistency with generally acknowledged fact.

♣ You said earlier that the superiority of MLET was not based on an improved ability to predict, but on better understanding. Now you claim superior ability to predict.

→ I said in few cases does the superiority of MLET arise from improved ability to predict direct measurements — in many cases it simply provides a sensible base for derivation of the correct mathematics. One hundred years of good experiment has assured that the current mathematics, when applied according to well defined procedures, will predict the observed measured values. I said in few cases does MLET do better, but I would argue that, in almost every case that goes beyond current knowledge, where concepts have not yet been tested, MLET does make the superior prediction.

♣ For instance?

→ The gravitational wave prediction for one.

♣ But you don't have any results, so that claim isn't very convincing.

→ OK, let's look back to 1913. SRT would never have predicted the Sagnac effect. MLET clearly required it.

♣ Whatever. . . In any case, if I understand you correctly, MLET would be completely discredited by the unequivocal demonstration of gravitational waves (of the expected form, of course!).

→ That's right. MLET's superior ability to predict clearly tells us that they don't exist.

♣ And if they are found to exist, you'll admit you're wrong, shut down your project and put the remaining money to better use. Right?

→ In your dreams. Sure, the credibility of MLET is on the line. But don't make the mistake of betting against us — you'll lose. *Gravitational waves of the expected form will never be detected.*

♣ After all this, you still expect me back here tomorrow morning?

→ Same time, same place. Have a nice day.
— *** —

The Tahoe Project was conceived by Steven Mitchell and funded by Jim Price.

As many of you probably know, Price began as an electronic engineer in the early sixties, and through the success of his own inventions and several start-ups that paid off unbelievably well, he is currently a billionaire several times over. He has established the non-profit RealityCheck Foundation with initial funding of three-hundred-million dollars. Fifty million of this has been allocated to the Tahoe Project.

About the Tahoe Project.

Steven Mitchell heads the project.

I'm Director of Public Relations, a rather pretentious title at this stage, but with fascinating (to me, at least) responsibilities. My name is Julia Clark.

The project is structured to support MLET in three ways:

1) Explore and document theoretical implications.

2) Establish credibility within the physics community.

3) Develop teaching materials and establish a formal research/teaching program for interested qualified students..

Steve, though General Director, will concentrate on the first point. He's hired three recent graduates from strong programs: Paul Donovan from M.I.T., Sarah Fishburn from Stanford, and Tim Hathaway from Caltech to assist with theoretical work. They are supported by two computer-graphics specialists and a professional science writer.

I've been given a wonderfully free hand to explore methods of developing credibility (the second point). This book describes some of my initial efforts.

Over half of the funding is expected to be allocated to the third point. Specific planning in this area is expected to lag the other efforts by six months to a year because of the needed support from the first two efforts.

Back to my effort.

Price has high hopes of gaining some acceptance within the physics community within the next few years. Mitchell doesn't.

Steven Mitchell believes that it will probably take decades even if MLET meets every challenge. He feels that (in essential agreement with Thomas Kuhn's, *The Structure of Scientific Revolutions*) very significant doubt must precede any toleration of serious questioning of the very strong current consensus. He argues that, while very serious problems with current theory are acknowledged, there is little evidence of any ability to question the current worldview.

So Price expects some possibility of early inroads while Mitchell doesn't. I'd like to prove Price right. What we're trying to determine in these early efforts is how best to gain a serious hearing. Price has agreed to cover the cost of my proposed approach.

We aren't the only people who disagree with current theory. There's a lot of good work being done but there's also a very large flood of nonsense. Obviously it won't be easy for us to get a serious hearing in the current environment. I argued for setting up a specific target — pick one very credible physicist and do whatever it takes to get him to listen. Ideally, make him want to listen, but failing that, nevertheless make sure he *does* listen. To do that we needed to make an offer he couldn't refuse.

Again, I wanted a physicist with very solid credentials — credibility within the physics community was of primary importance. Secondly, I felt we should go with youth. It seemed likely that any older physicist, with possibly decades already spent working with current theory (work that very well might be significantly invalidated by the changed perspective) might naturally be very resistant to the change of perspective.

We did a lot of research, narrowed our interest down to seven people, did additional research, talked to colleagues of the seven and narrowed our list to three outstanding young physicists. Price has excellent contacts within the physics community so we asked him to narrow our choice down to one person. He used his contacts, made his choice and got the commitment we were after from David Rhodes.

I was later to learn that he had told David nothing very specific. All David knew was that the billionaire engineering and business genius was asking for one hour a day of his time for thirty days. In return Price would pay him thirty-thousand dollars plus all expenses for a thirty-day stay in a luxury condominium at a resort near the south end of Lake Tahoe.

Price had moved fast and we only had two weeks to get ready for David's arrival.

Price and Mitchell had previously agreed with my suggested pacing. Mitchell would present our ideas to David in thirty one-hour-maximum sessions, one each morning for thirty days. These would, as far as possible, be structured as conversations but with Mitchell rather rigidly controlling the subject matter and pace. David had agreed to allow us to video tape each session.

Mitchell could have easily covered everything he wanted to cover in three or four full days, but we all felt that any possible impression made in such a short time would dissipate just as quickly. This way David would be forced to think of MLET at least thirty times over thirty days. Also, by scheduling the talks between seven and eight o'clock each morning, we minimized the negative impact on both David's vacation time and Mitchell's primary commitments.

We redesigned the living room of a top floor suite to serve as our 'interrogation' room. The room had a wonderful view of the lake, affording a quiet, peaceful setting. With a one-way mirror installed between it and my adjacent office (a converted bedroom) the 'interrogation' name seemed appropriate. I controlled a video camera from the office behind the mirror and so was able to watch and listen to the conversations without intruding.

So I've recorded the conversations. This book is primarily a simple transcript of those conversations, hence the 'Conversations at Tahoe.' The conversations frequently reflect or refer to the content of various papers. Some of these I had planned to include in an appendix.

The opening conversation regarding LIGO and the various other gravitational wave detection efforts is from the first conversation between Steven Mitchell and David Rhodes and took place in the interrogation room of the Tahoe Project building. This building is a small part of a larger resort that Price recently developed. Our building is four stories high, with thirty separate condos, a cafeteria, various conference rooms, work spaces, and offices.

Almost all of the conversations recorded here were between David and Steve. The first morning David had arrived right at seven, breakfast for both had been brought in from the cafeteria and the session began immediately. When it was over Steve brought David to my office as I shut off the camera and labeled and stored the morning's tape. Before he left us Steve suggested that I show David around the facilities.

As I cleared my desk I noticed David watching me. In response to my "What?" he replied, "You have a bachelor's degree in physics so you know as well as I do that he's nuts. How'd you ever get involved with these people?"

I rather enjoyed his frank incredulity. "I chose to," I replied, "for a number of reasons. Ask me again at the end of your thirty days — you'll be better able to understand then."

"Afraid your reasons would reveal questionable ethics?" David asked, "I'd bet the money had a lot to do with it. Price's very generous. Just plain throwing his money away if you ask me."

"He believes in the project. And yes, he's generous with those he thinks key to success. He wants to be sure that nothing stands in the way of strong commitment."

"Why go through all this with me? With his money and reputation you could buy a great deal of publicity, if that's what you're after."

"Publicity is not credibility." I replied. "And publicity without credibility would only build resentment. Even with lots of money and exceptionally good theory, credibility won't come easy — and is virtually impossible without a hearing. You're here as part of an experiment to test how best to get a serious hearing."

"You're wasting your time, you should know that."

"Mitchell would probably agree with you. For all of us, it's an experiment. The extent to which we might influence your thinking will tell us a little about what we might expect from the physics community as a whole — this is a first step for us. Since your personal credibility is exceptional for someone so young, you were an obvious choice. So we made you an offer you didn't refuse. And here you are."

"He's allocated fifty-million dollars to your project. Just imagine what sensible people could do with that kind of money. And why is Price supporting this anyway? Surely he must realize that if the ideas were sound there wouldn't be any need for private funding; every university in the world would be scrambling to include it in their research, if not in their curriculum," David said.

"Get real."

"You're right, the ideas are so far out that I'd bet nobody even tried to get mainstream support before Price decided to throw his money away."

"No, nobody did — for many reasons."

"Such as?"

"For one thing, we all felt that proper respect for consensus demands that efforts such as ours be funded outside normal channels."

"Why?"

"Science can be severely compromised by unseemly support of iconoclastic ideas. We're outside the pale. Given the strength of the current consensus we *should be* outside the pale. Unless and until we can change consensus thinking, mainstream scientific organizations have no business supporting our effort — tolerating, maybe, actively supporting, no. Of course we want to change the consensus, but until we do we'll stick with private funding."

"An interesting outlook. If you know you're right, it's hard to see how science would be compromised by community support."

"In a sense you're right — on the rare occasion that the iconoclast is right physics would benefit from supporting the new concept. Unfortunately, though, any general inclination to support iconoclastic ideas would probably do more harm than good."

"I still think that's weird. You don't want mainstream support?"

"It's not that we don't want it," I replied. "How can I

explain? Look, I think it appropriate that, in order to maintain public confidence in any scientific discipline, there be 'protectors of the flag,' guardians of the consensus if you will, that stand firm against iconoclastic claims. At the same time, I think it's appropriate that individuals making up that consensus maintain a healthy skepticism.

"In our case, given the very strong consensus against us, support from the physics community would surely be seen by most as a cynical scrap thrown to undeserving dogs; and would probably hurt more than help — every kook in the world would demand similar consideration. We'll stick with private support, thank you."

"Don't get me wrong," David said, "I certainly don't think you should get any support for promotion of such crazy ideas. I only pointed out that if your concepts were at all plausible you could easily find support. Not only do you not have any chance of support, you don't even seem inclined to welcome knowledgeable criticism. I get the impression that Mitchell isn't much interested in my opinion, he's simply telling me the way it is."

"You're right. In fact, none of us are much interested in your opinion at this point," I said. "We've collectively spent years looking at possible objections to MLET and we understand the consensus viewpoint. Our real interest is whether or not, or to what extent, your opinion can be changed. We think the arguments in favor of MLET are convincing, so you're being paid to listen, not to enlighten."

"Don't you think that a bit ridiculous?" he said. "You yourself said my credibility was exceptional. So why not listen to me?"

"I'd very much like to have this conversation again in about a month," I replied. "As I said, we think we understand your position rather well. This month is designed to help you understand ours."

"It seems rather insulting. I'm supposed to just sit and listen, with no opportunity to help you understand how wrong you really are."

"Oh, you'll have your chance," I assured him. "You're free

to take notes and spend as much time as you like thinking about our claims. And at the end of the month we'll be very interested in what you think."

"But with no chance of changing anyone's mind?"

"Zero, zilch, none," I laughed. "Seriously, though, at the end of the month we will be interested in what you think."

"So you're as sure that MLET is right as Mitchell is?" he asked.

"Yes."

"How did that happen?"

"I can read. I can think."

"Reading and thinking led you to believe MLET is superior to current theory? You're every bit as insane as Mitchell. You do know that MLET requires an ether?"

"Look, let's leave this discussion to you and Mitchell."

"Why? Are you afraid that I'll make you doubt MLET."

"No, but realize that the time we spend in this kind of discussion doesn't count toward your hour a day. You'll still owe Mitchell his hour."

"I understand that. But that doesn't mean that challenging you is off limits, does it?"

"No, of course not. Just off the clock."

"Then let me ask again. Why do you prefer MLET?"

He knew that to answer that completely I would have to use all the arguments that Mitchell meant to use over the next thirty days. I refused to be drawn in. "Ask that again at the end of the month."

Note that in the transcripts:
- ♣ David's comments always begin with this format and
- → Mitchell's with this.

Note also that the Modified Lorentz Ether Theory is abbreviated MLET and Einstein's Special Relativity Theory is abbreviated SRT throughout.

Challenging Consensus
[Steven Mitchell and David Rhodes]

11 May, 1999

→ Since I've promised to meet a friend from the Bay Area as soon as he arrives, I may have to leave at any time, so I'd like to stick with some very general observations this morning

♣ You're the boss.

→ OK. Let's get to it.

You've met Jim Price. Jim thinks that if MLET is as good as we think, then it should achieve general acceptance in just a few years. I don't agree. Even adding null LIGO results to proofs of superiority, I think that the change is too iconoclastic to make much of an impression in the immediate future. I don't doubt that the essential MLET perspective will eventually be accepted, I just think that the current consensus can't tolerate such a radical change.

In general, what do you think of the physics community's openness to such change?

♣ It'd depend entirely on the strength of the evidence in favor of the change. There's a tendency to blame lack of acceptance of radical change on lack of openness within the community, when the reality is that rejected claims are almost always simply wrong.

→ There's certainly a lot of ridiculous claims being made by people who should know better, I'll not deny that.

♣ I'd have to say that your claims are right up there with the worst of them.

→ And, understandably, people aren't very open to ridiculous change. That's my position. Evidence in favor of iconoclastic change will never be sufficient in itself to effect change. A weakening of consensus must come first. As long as no doubt is allowed, no contradicting evidence can be entertained.

♣ If I'm understanding you correctly, you're saying that the physics community is unreasonably closed to new ideas.

→ Unreasonably closed to new ideas? No, that's not what I'm saying at all. But closed to iconoclastic ideas? Certainly. And not just physics. I would argue that any scientific discipline is pretty securely closed to iconoclastic concepts.

♣ You're wrong. We're often accused of being rigid in our positions — but that's grossly unfair. If there were meaningful evidence that MLET was right, then Price would have every reason to be optimistic. Notice, I said, if there were meaningful evidence. I'd guess that you gloss over the weakness of your position and simply blame the consensus community for not seeing the evidence.

→ Not at all. At this point I think I understand their inability to see the evidence better than you do.

♣ Yeah, right. I'm reminded of something Leon Lederman[10] wrote in his book, *The God Particle*:

> . . . what is rarely understood by the lay public is how ready, how eager, how desperately the collective science community in a given discipline welcomes the intellectual iconoclast — if he or she has the goods.

Lederman's right. The physics community has frequently had to deal with, and so is necessarily very open to, even iconoclastic change.

→ I hope you won't take offense at my amusement. Remember your reaction to the MLET requirement for an ether?

♣ The possibility of the existence of an ether was disposed of long ago. It's not something open to question.

→ That's my point. Where there's no doubt —

♣ We know for a fact that there's no ether.

→ Exactly. And when 'we know for a fact,' when no doubt is allowed, there's no openness. That's all I'm saying. I agree that your position is reasonable — provided only that you're absolutely certain that no ether exists.

You just did something that most physicists (most scientists, even most reasonable people, for that matter) do. You overlook significant implications of Lederman's *if he or she has the goods.* It's a given that the iconoclast must demonstrate the reality of the goods to make any impression at all. Right?

♣ Of course.

→ Sensible. Reasonable.

But that leads directly to my point. The impact of this requirement for the goods, with regard to the overall validity of the claim of open-mindedness, is grossly underestimated. Even Lederman doesn't seem to fully recognize the implications of that very reasonable qualification.

♣ What do you mean?

→ Look. In my dictionary, iconoclast is defined as *one who attacks settled beliefs.* And isn't the phrase 'one's settled belief' pretty much synonymous with the phrase 'one's truth.' An iconoclast's supposed goods must therefore, by definition, conflict with the community's unquestionable (how else, settled?) truth, and so, unquestionably, the iconoclast's goods must be deemed false, clearly unacceptable. How then, can any scientific community possibly be expected to welcome the claims of a true iconoclast? These people aren't idiots.

♣ You're playing with words.

→ I'm playing with a rarely acknowledged reality.

Again, by definition, a true iconoclast's claims must contradict accepted truth. Clearly, in the face of sure knowledge (settled belief) to entertain iconoclastic change is to entertain contradiction — and where does the motivation for that come from?

♣ And your point is?

→ Specifically, MLET is unambiguously based on an ether, and an ether requirement simply rules out any credible consideration of MLET's goods. Don't you agree?

♣ Probably.

→ What do you mean, probably? Are you conceding that there just might be an ether?

♣ No way. As I said before, we know better.

→ So you persist in making my point for me.

As long as you know that you know better, you won't willingly waste your time on such iconoclastic claims. Doubt, a weakening of consensus, a 'probably not' in place of a 'No way!' a suspicion that you *might not know for certain*, must occur before iconoclastic change can be entertained.

And where can such doubt come from?

♣ From convincing evidence. That's the bottom line.

→ To use your eloquent expression — yeah, sure. MLET

faces an overwhelming contrary consensus — not only does it embrace an ether, but — heaven help us — it contradicts SRT. Is that iconoclastic enough for you? Are you ready now to examine our 'goods,' our convincing evidence?

♣ Physicists aren't idiots. Of course there's no openness to such nonsense.

→ So what about Lederman's claim, "If he or she has the goods"?

Without going any farther, without any pretense of examining any claim of supporting evidence, you wouldn't hesitate to declare MLET nonsense, and turn away with a clear conscience. Isn't that right?

♣ Of course. But it's not really lack of openness, it's simply lack of any inclination to waste time. I'd rather spend my time on something with better odds.

→ It amounts to the same thing — you're not inclined to waste your time examining claims that you know, up front, contradict what you know to be true. Further, wouldn't you expect the same reaction from any credible physicist?

♣ Definitely.

→ So we agree.

Again, you people aren't idiots. Given your settled beliefs, you're behaving quite sensibly. I'm not trying to gainsay that.

♣ What's so funny?

→ I just happened to think of something I saw on TV a few weeks ago, and it suddenly occurred to me that it was a perfect parody of someone who is open to iconoclastic change.

I was reading with the *Dharma and Greg* show in

the background, only half following the action, only vaguely aware that accusations of closed-mindedness were being bandied back and forth, when Dharma gave Gregg her definition of open-mindedness, "You know you're open-minded when you can embrace an idea that you *know* is just plain s t u p i d !"

Again, openness to iconoclastic change is openness to the contradiction of what one knows to be true (settled belief).

♣ I suppose that may be true, if one accepts your definition of iconoclastic.

→ Mine? Webster's.

If you and Lederman are using some other definition then you should at least make it clear that you're doing so. Given the accepted definition of iconoclastic, the supposed openness to iconoclastic change is an illusion. Openness to contradiction of one's truth simply doesn't make sense. How, can we possibly ask anyone to set aside Einstein's worldview — when they know beyond any possible doubt that it's right? Surely, we can't ask people to question that which they know with absolute certainty?

♣ We know, sure. But 'with absolute certainty'? No one claims absolute certainty.

→ No? Look at what's actually being said.

Carl Lanczos[11], in a lecture titled *The Greatness of Albert Einstein*, delivered at the University of Michigan in the spring of 1962 as part of a lecture series titled *The Place of Albert Einstein in the History of Physics* gave us these priceless observations:

> Nobody intends to diminish the merits of other great men of science, but there was something in Einstein's mental make-up which distinguished him as a personality without peers. He wrote his name in the annals of science with indelible ink which

> will not fade as long as men live on earth. There is a finality about his discoveries which cannot be shaken. Theories come, theories go. Einstein did more than formulate theories. He listened with supreme devotion to the silent voices of the universe and wrote down their message with unfailing certainty.
>
> What was so astonishing in his manner of thinking was that he could discover the underlying principle of a physical situation, undeceived by the details, and penetrate straight down to the very core of the problem. Thus he was never deceived by appearances and his findings had to be acknowledged as irrefutable.

"Unfailing certainty" and "irrefutable." Where's the room for any possible doubt?

And Lanczos is hardly alone. Nigel Calder[12] writes:

> Einstein's theories are the bedrock. . . . It *is* Einstein's universe.

Paul Davies and John Gribbin[13]:

> All of the implications of special relativity . . . have been confirmed by direct experiments. There are still people who believe that it is all "just a theory" . . . but *they are wrong*.

And John Gribbin[14] claims that Einstein's concepts are:

> . . . fully accepted by all except a tiny minority equivalent to the flat-Earthers, who still don't believe the Earth is round .

Isaac Asimov[15]:

> No physicist who is even marginally sane doubts the validity of special relativity

Clifford Will[16]:

> Special relativity is so much a part not only of physics but of everyday life, that it is no longer appropriate to view it as the special "theory" of relativity. *It is a fact* . . .

What do you think? Is there anything here that might be seen as a concession that the knowledge is less than absolute?

♣ We're pretty confident, sure.

→ Pretty confident? *Unfailing certainty? Irrefutable? No physicist who is even marginally sane doubts the validity?*

In the eyes of the consensus community we're clearly 'less than marginally sane.' But there's worse. Paul Davies[17], writes of Herbert Dingle:

> In his later years, Dingle began seriously to doubt Einstein's concept of time. He had little difficulty persuading a motley group of followers of the absurdity of relative time . . . the mood of dissent he championed lives on, widespread and festering. I wonder why? Einstein must have touched a raw nerve.

So now we're a motley group of followers of less than marginally sane crackpots.

Davies is right in his suspicion that a raw nerve may have been touched. Dingle[18] wrote:

> It is simply that physicists have, unawares, allowed their trust in special relativity to escape the control of reason and become a blind slavery to dogma . . .

Dingle was at one time a respected member of the consensus community — had written a book explaining special relativity — and I'd bet that until he himself had come to doubt, he'd believed that there was within the community an openness to even iconoclastic change if only one had the goods. He quickly learned otherwise. Granted, he shouldn't have been so surprised. To any who have closely observed the effects of strong consensus his surprise and rage at his treatment does seem a bit naive. However, Davies' intimation that it was special relativity itself that 'touched a raw nerve' is disingenuous — SRT is wrong — that's not particularly bothersome. No, it's the total inability to allow questioning — it's the arrogant contempt showered upon any who dares to question — it's the absolute, unquestioning and unquestionable adherence to SRT, that so shocks and, yes, sometimes enrages.

Again, given the absolute surety of the consensus community, it's understandable that they wouldn't give anyone who questions SRT, a hearing — we accept that. But the hard part, the frustrating part, the maddening part, is the profound difficulty in understanding where this surety comes from. All experiments better support the clear alternative to SRT's explanation of relativity effects.

♣ You persist in making ridiculous claims with no supporting evidence. Show me this better explanation.

→ The evidence involves disagreement with SRT, and requires an ether. Can you get past either of those points?

♣ Me? You're dead. There's no way you can get past your rejection of well known facts.

→ It's a chicken and egg issue — without some doubt with regard to 'well known facts' there can be no hearing, and without a hearing there's a very minimal opportunity to introduce evidence, to effect doubt. SRT is flawed — more than that, it's just plain wrong — the world around us clearly shows that. But —

♣ Absurd claims are no argument at all. Give it up.

→ The claims are remarkably easy to support — that's what drives one up the wall when all refuse to hear. In any case, we're very aware that we're outside the pale — among Davies' despised motley group. So bear with me — I will, over the next few weeks, be making claims that you, with absolute (almost?) certainty, believe to be false. That's unavoidable. I'm as sure that you're wrong as you are that you're right.

So, with nothing to lose, I may, at times, be deliberately provocative, arrogant and obnoxious — in keeping, perhaps, with the smug, supercilious contempt for the skeptic that so enraged Dingle. The frustration is not significantly lessened by the fact that, if I knew for certain SRT was right, I'd behave similarly. Again, the frustration remains because it's extremely difficult to understand the pervasive lack of doubt, given so many reasons to doubt. Your position is weak, the refusal to question little short of criminal.

- ♣ Who do you think you are? Einstein offends your intuition and so you're gonna show us all up for quacks.
- → And you know, without even a glance at the goods, that we're wrong. Not only are we insane, not only are we challenging absolute fact, but we're only one small group among a great many kooks out there; only one insignificant part of Davies' *motley group*. Without culture, we simply can't accept irrefutable truth — we're poor, ignorant, modern-day flat-earthers.
- ♣ Believe it.
- → And that's not the whole of it. A reasonable person must ask: "What motivates such insanity?" Why do we persist in challenging Einstein? John D. Barrow[19] writes in *Theories of Everything*:

   . . . scientists . . . receive a large amount of mail from misguided members of the public announcing the discovery of their new "Theory of the Universe" (the author has received two during the last week alone). . . . They aim to show how Einstein was wrong in some way and . . . have an obvious psychological motivation. Einstein is perceived as the twentieth-century scientist *par excellence*, and hence it is fondly imagined that, by catching him out on some point, the new author would be hailed as the new scientific messiah, greater than Einstein.

No matter. So be it. We knew from the start that if one can't put aside all skepticism with regard to the certainty of SRT, then theoretical physics is not a field offering one any hope of respect.

♣ No one wants to hear your screwy ideas so you think you're being treated unfairly. Face it — some facts are just that — proven facts. There are, whether you like it or not, cases where there's clearly nothing to gain by listening to patent nonsense.

→ Of course. And again, bottom line — no doubt, no hearing.

♣ Colossal ignorance isn't appealing.

→ Of course not.

But tell me something. If, like us, you believed in concepts known to be iconoclastic with respect to such a solid consensus, what would you do?

♣ Seriously? Have my head examined.

→ It seems to me one has two possible options: (1) Wait for a growing confusion and a multitude of problems to weaken the strength of the consensus, a process that is happening and will ultimately succeed (although it may take many decades); or (2) Work to create doubt, show the superior alternative.

♣ Reject SRT and re-introduce an ether? Neither path has a prayer.

→ The first option would seem attractive except that it'll probably be an excruciatingly slow process due to the natural instinct to set aside confusion in favor of ad hoc adjustments to the old theoretical base. This generally means that growing confusion tends to be dismissed as a growing understanding of just how complex nature really must be. Eventually, this process will be recognized as the failure it is, but tortured accommodation has been embraced for decades now, and the intractable nature of the many problems may well go unacknowledged for many decades to come.

It's probable that the current elite of the modern physics community must die off before doubt (doubt sufficient to allow consideration of MLET's iconoclastic concepts) can overthrow the current firm consensus. Julia and Jim think that significant early inroads of doubt might, just might, be made among the younger members of the elite. Perhaps they're right. I have serious doubts even there.

♣ I'm sure your worst fears will be realized.

→ But, I'll admit, I was quite pleasantly surprised that we were able to attract three very bright students, each from a very strong program, to work with us. (Just don't ask what incentives Price offered!) So even with your advanced degrees, I'd have to admit, you do have youth in your favor.

♣ An amusing analysis. So what do you really expect of me?

→ Even for the young, doubt will come hard. And you have your Ph.D., have done postdoc research (exceptional work, I hear) at both SLAC and CERN — your membership in the elite community is secure — so for you, doubt surely won't come easy. Still, you're only twenty-nine years old!

♣ So I'm here as the subject of an experiment?

→ That's right. But you already knew that. Julia wants to find a way to get us a hearing, so she's bought some of your time. Can we translate that into some small semblance of doubt? Can we grow doubt to the point where the possibility of an alternative can be entertained? Finally, can we make a convincing case for an alternative?

♣ A ridiculous waste of time and money.

→ I think so, but Julia wants to do it and Price's paying for it. So here we are.

Let's end today's discussion with some observations you might agree with. I think we can agree that some facts are pretty well proven, and no one has anything to gain from denying obvious truth.

♣ But isn't that exactly what you claimed? In order for iconoclastic concepts to be acceptable doubt must first be created with regard to one's truth? Yet you're conceding that the truths are not to be denied.

→ Cute. You know very well that 'one's truth' isn't used in the same sense as my 'obvious truth.' The difference is the difference between theory and raw experimental data. Prolonged acceptance of a theory and absolute disallowance of skepticism within the community, tends to create the impression that theory is fact, and so theory may very well become part of 'one's truth.' But you know as well as I that theory is never fact in the same sense as raw data.

It's the raw experimental data, the data that's the foundation of modern physics, that's simply too good to be seriously questioned. Too many experimenters have done too good a job, have provided too solid a base of objective fact, to allow any claim of serious error here.

♣ So where're you going with this? It's this very data that proves SRT is correct.

→ Facts prove theory? Come on. You're a better scientist than that. These particular, universally acknowledged

facts do not, and never can, prove that SRT is fact. Again, you know as well as I that theory is never simply fact, never really proved. A future uncovering of contrary fact or the emergence of a better theory can never be completely ruled out. Stephen Hawking[20] reminds us that:
> Any physical theory is provisional, in the sense that it is only a hypothesis: you can never prove it.

♣ The point's moot. You're still going against known truth.

→ Your opinion. The truth is (Davies and others notwithstanding) SRT is *not* fact, it remains theory. And, contrary to common perception, it's not a particularly good theory, not even very plausible when forced to deal with the totality of the experimental data.

♣ What in the world am I doing here!

→ In any case, I think we can agree that the data is very good and the mathematics is, at the very least, not seriously flawed.

♣ The current data and the current mathematical base? With respect to all of physics? Those come from current theory. What's your problem? — you can't possibly believe that the data and mathematics of current theory actually *favors* a radical new theory.

→ I do — it does. The data and mathematics do strongly favor MLET as the better theory. Clearly.

I've seen claims that the necessary mathematics dictates the acceptance of SRT. If we thought that were true, we would say "So be it" and join the consensus. But, as Richard Feynman[21] warned: *"Many physical pictures can give the same equations."* And others have reminded us that actual experience has shown that one can sometimes come up with the right mathematics from very erroneous models. Not only is a different physical picture possible, but it's better supported by the data and the current mathematics than is SRT.

The strong current consensus tells us that physicists think that, in SRT, they have the best possible explanation of the effects of velocity. Let's be honest. The problem is not that they *think* they have the best explanation possible — the *real* problem is that they *know* that they do.

And how do they know? Well, for one thing, they know that Einstein, in the words of Carl Lanczos, "listened with supreme devotion to the silent voices of the universe and wrote down their message with unfailing certainty."

Just as a matter of curiosity, do you think there is a significant difference between the phrases *absolute certainty* and *unfailing certainty*?

♣ I imagine Lanczos might be uncomfortable with 'absolute', but I'd be hard pressed to grant much difference.

→ So you'll concede that we're not being unduly extreme when we claim that the consensus community does insist that SRT is absolute fact?

♣ I do object to your use of the term 'absolute.'

→ What's the difference between 'fact' and 'absolute fact' in this context?

♣ Whatever the difference, I will concede that we're certainly very sure that SRT is fact.

→ But not absolutely sure?

♣ Not absolutely sure, no.

→ All I ask is that you remember that over the coming weeks. You'll still have trouble, because you are 'very sure.' But if you can set aside the 'absolute' for these few weeks we'll show you that we can do better.

♣ You're sure about that?

→ Absolutely!

♣ So the entire physics community is wrong? A bit arrogant, don't you think?

→ Arrogant? Tell me about arrogance. *"No physicist who is even marginally sane doubts the validity of special relativity."*

♣ But you must admit that when you're right, strong statements aren't terribly objectionable. Strong statements are justified in support of excellent theory.

→ Sure. But how do we identify excellence? Current theory struggles with some very significant anomalies, and there *is* a serious level of confusion. Might current theory be thought less than excellent?

♣ Nobody has ever done better.

→ The current surety precludes anyone within the culture even attempting to do better. I'd argue that your 'nobody' applies only to those within the consensus community. Others have done better in many specific instances. But let's leave that for now. I'd like to go back to the point I was trying to make earlier.

Given that the existing empirical and mathematical base is too good to allow error in these areas of sufficient magnitude and scope to explain the current confusion, where do the problems come from?

It seems to me that sensible people, even if they weren't personally aware of a better worldview, might reasonably question the supposed excellence of the current paradigm. Sensible people would, should, must, ask "What is the source of this confusion?" "Where can we hope to find the confounding error?"

And, "Why the intractable domain boundaries?"

♣ There may be several reasonable answers to those questions. Have you ever considered that maybe we just don't yet know enough? Or perhaps the confusion arises because we can never know? Perhaps "that's just the way Nature is." Many laymen are simply unable to appreciate the complexity and non-intuitive nature of the full implications of SRT.

→ You can bet your life that as long as you hold to SRT you'll be forced to also hold to 'that's just the way Nature is.' You'll never, within SRT constraints, get beyond the current confusion.

But "not appreciate the implications of SRT"? Who're you kidding? It's those very implications, the unavoidable lack of consistency, coherence and convergence, that's so damning. If we're not content with 'that's just the way Nature is,' we'd better be interested in where the intractability comes from.

♣ And you've discovered a new interpretation that solves these problems?

→ At the very least we've found one vastly superior to the current consensus. And it's a worldview that's strikingly unambiguous in predicting null LIGO results.

♣ So we'll soon have proof that you're wrong. Then what?

→ Dream on.

♣ You've used the term 'worldview' — that sounds rather pretentious. You have a radical perspective, but worldview?

→ Given one conservative theory versus one radical theory, the difference is radical. But blame that on Einstein. MLET is by far the more conservative.

But why 'worldview'? Can you suggest a more

appropriate word? Einstein said that the most important thing that SRT revealed was that the requirement for absolute symmetry of all of the laws of physics, with regard to any possible inertial frame and at every level of abstraction, was a good test of the validity of all subsequent theory. His symmetry is an absolute, and so there can be no true ether frame, no frame physically different than any other.

♣ And you disagree? You're wrong.

→ Whether we can identify such a frame by direct measurement is really beside the point; sensible understanding requires that it, an ether frame, be real.

♣ You're beating a dead horse.

→ Oh, really?

Sensible understanding doesn't require ability to directly detect the ether frame, but the fact that we *can* detect a 'background radiation' inertial frame, and so measure velocity relative to it, must at least give an interesting twist to your 'dead horse' assertion.

What can you say about the fact that Doppler measurements give a clear velocity relative to this universe-wide background radiation?

♣ It's interesting, but certainly no problem.

→ Of course it's not. Since it's just another inertial frame, it doesn't directly affect predictions that can be measured, so, to the true believer, it's a metaphysical question.

But we like understanding as well, and it seems obvious that, with light being gravitokinetic oscillations within the ether, the background radiation frame might very well be the elusive ether frame.

♣ So you think that when we measure velocity relative to the background radiation, we are in fact measuring velocity relative to the ether itself?

→ Whether we are or not, whether the ether frame is directly detectable or not, sensible understanding demands that there *be* a real ether frame. And if the ether frame is real, it seems to us that we should expect the background radiation to reflect this frame.

But let's get back to 'why worldview' considerations.

Notice that MLET requires the same measured (apparent) symmetry that SRT does, but asserts that the measured values themselves require the reality of an underlying, sensible asymmetry. In contrast, Einstein clearly intended to establish his SRT absolute, his absolute symmetry, the physical equivalence of all inertial frames, as an over-arching principle. With his success his 'perfect symmetry' absolute has become the foundation of the prevailing worldview, if you will, of twentieth-century physics. As Nigel Calder[22] put it with respect to the consensus viewpoint, "This is Einstein's Universe."

♣ But Einstein's Universe embraces much more than just SRT. What about general relativity?

→ General relativity took the form that it did because of the constraints of SRT's absolute-symmetry worldview. MLET, freed of the constraints of SRT, incorporates a mathematics almost identical to general relativity, but couples that mathematics with a refreshingly sensible physical model.

In any case, Einstein's relativity insists on an absolute symmetry in every aspect of physics. MLET likes the symmetry, embraces the symmetry, but recognizes it as an apparent (directly measurable) symmetry derived from an underlying velocity induced real asymmetry. The measured symmetry is a wonderful phenomenon — the sensible effects of velocity relative to the ether. This apparent symmetry provides a very satisfying, universally consistent reflection of the underlying asymmetry.

So the differences between MLET and the current consensus are core differences — in one sense strikingly small (the *apparent* of MLET, versus the *absolute* of Einstein) but at the very foundation of physics — they are *worldview* differences. And that's what's needed.

♣ Since, as you've acknowledged, current theory works so well, why is there any need for such a drastic change of perspective?

→ Works well insofar as necessary mathematical procedures have been defined to accommodate the experimental data — but fails miserably with respect to any hope of understanding. Why should we ask for more? At the leading edge of physics, after nearly a century of SRT, increasing knowledge seems almost in lock-step with increasing confusion, a condition that I think is strikingly symptomatic of worldview problems — symptomatic of error at the very core of the fundamental perspective.

♣ You know that MLET doesn't have a prayer of ever being accepted, and you're driving me nuts with your perverse inability to accept that.

→ So maybe you can began to understand the source of Dingle's rage. You know you're right, just unable to get me to listen, unable to get a hearing. But you'll soon be back within the comforting arms of the consensus community. Now, if you were ever to question SRT —

♣ Your claims are outrageous. Certainly iconoclastic, but don't pretend that that's the reason you can't get a

hearing.
- → For now, I expect to be heard because we're paying you to listen. I'd be surprised if any hearing goes much beyond that in the near future. Until the physics community can entertain some tiny doubt with regard to SRT, I think it would be unreasonable for us to expect anything more.
- ♣ I'll listen for the agreed on hour a day, but don't expect me to do more than that. I'll say it again, I think your claims are absolute nonsense.
- → That's all I've ever expected.
- ♣ And I'm not rejecting your claims because I'm unwilling to question. I'm rejecting them because they're wrong — not because they're iconoclastic.
- → Nonsense. The concepts that make them wrong are what makes them iconoclastic — you know they're wrong precisely because they're iconoclastic (contradict settled belief). All that we've disclosed is that it challenges your certainty that SRT is right — MLET challenges settled belief. That is the whole reason that you're so sure that we'll never change your mind.
- ♣ OK. You're right. I'm rejecting your claims because they contradict settled belief. But I know why I believe as I do. I know that I'm well educated. I know that I understand modern physics far better than you do.

  I'm tired of this discussion. Like Dingle, you're blaming lack of openness when the real problem is your own ignorance. I'm outa here.
- → It has been a little over an hour. Sorry about that. But before you go, let me try once more to make the point I was aiming for earlier. We start with the consensus that current theory has significant problems in many areas.

- ♣ Problems, yes — but not with respect to SRT.
- → God forbid — of course not! But then where *does* the intractability come from? Why the unyielding domain boundaries? We know that:
  - The *experimental base*, the experimental data, is *very good*.
  - The generally accepted *mathematics* is also clearly *very good*.
  - The *current worldview*? What can we say? After nearly a century, we are faced with disparate, non-convergent (if not completely incompatible) domains — Special Relativity, General Relativity, quantum theory.

Forgetting MLET for the moment, where would you look for improvement?

— *** —

As I watched him move toward my office I was sure that David was regretting having committed himself to listen to such 'nonsense' for another twenty-eight days. I was also sure that he was placing much of the blame for his current situation on me.

"How can you be part of this? What do you think science is — some game where consensus, where sensible rules don't matter? Modern physics is the foundation of all of modern science. Don't you understand the role of disciplined consensus in separating truth from fiction? Physics is a science, based on fact, not on some fanciful notion as to what *ought* to be."

"Disciplined?" I repeated. "So you do prefer discipline? You know, it's the imposition of arbitrary constraints coupled with the rejection of reasonable intellectual discipline, that makes current theory so flexible that any conceivable experimental result can be accommodated, given only moderate ingenuity."

"What do you mean by that ridiculous statement?"

"Look," I said, "I know that tomorrow Steve intends to try to establish a consensus definition of what good science is, of what constitutes good theory, and I don't want to anticipate him. But we can show that, by any measure, MLET exemplifies the superior discipline. And the lack of sensible discipline in current physics can be traced directly back to what we only half jokingly refer to as 'Einstein's Camel'."

"Einstein's Camel? Give me a break!" David laughed. "OK, I'll bite, what do you mean by that?"

"I'm sure you're familiar with the ancient observation that certain people may sometimes "strain at a gnat and swallow a camel." Because Einstein strained at a gnat, couldn't swallow one simple violation of positivistic dogma, modern physics has been forced to swallow a very large camel."

"How so?"

"Einstein's[23] inability to swallow the gnat," I explained, "was expressed in a 1920 speech in which he stated:
> For the theoretician such an asymmetry in the theoretical structure, with no corresponding asymmetry in the system of experience, is intolerable.

"What's wrong with that?" David asked. "The statement exhibits an essential discipline that you claim is missing from current theory."

"The gnat that Einstein couldn't swallow is Lorentz's underlying asymmetry," I replied. "Einstein found it intolerable, not because it didn't make good sense, but because positivism couldn't allow it."

"And the camel?"

"Swallowing the camel," I said, "is the unavoidable alternative to the underlying asymmetry. If the asymmetry is rejected, then one is left with no choice — the camel must be swallowed."

"And what does that mean?"

"Einstein's camel is his absolute — absolute symmetry with respect to any inertial frame at all levels of reality."

"And this symmetry requirement has damaged physics?"

"Severely damaged understanding — definitely," I answered. "It's a little difficult to encompass all of the negative implications of swallowing the camel in a brief summary. But three statements, taken together, may come close — one from Richard Feynman, one from Leon Lederman and one from Stephen Hawking.

"Feynman[24] wrote:
> The more you see how strangely Nature behaves, the harder it is to make a model that explains how even the simplest phenomena actually work. So theoretical physics has given up on that.

"And Lederman[25] tells us that:
> All we can ask of a theory is to predict the results of events that can be measured.

"And Hawking[26] seems to sum these two statements when he writes:
> I take the . . . viewpoint that a physical theory is just a mathematical model and that it is meaningless to ask whether it corresponds to reality. All that one can ask is that its predictions should be in agreement with observation.

"Together these statements describe the inevitable, destructive impact of Einstein's inability to tolerate asymmetry. The consensus community, in following Einstein's lead, has swallowed the camel."

"You people are crazy. What did I do to deserve this?'

"You think we're hard to take?" I asked. "Let me read two passages, one from Davies' and Gribbin's *The Matter Myth* and the other from Kip Thorne's *Black Holes & Time Warps*. First Davies and Gribbin[27]:
> I believe that the reality exposed by modern physics is fundamentally alien to the human mind, and defies all power of direct visualization. . . .

And
> The realization that not everything that is so in the world can be grasped by the human imagination is tremendously liberating. . .

"I have no problem with any of that," David insisted.

"You've swallowed the camel. And without the underlying asymmetry, concepts alien to the disciplined mind are unavoidable. Davies' 'liberating' is in fact a direct acknowledgement of the destruction of essential discipline.

"Let's look at Thorne's discussion regarding some general relativity concepts. After describing how a flat space-time with gravity affecting both perfect rulers and perfect clocks could explain the same phenomena explained by general relativity's curved space-time, Thorne[28] wrote:

> What is the real, genuine truth? Is space-time really flat, as the above paragraphs suggest, or is it really curved? To a physicist like me this is an uninteresting question because it has no physical consequences. Both viewpoints, curved space-time and flat, give precisely the same predictions for any measurement performed with perfect rulers and clocks, and also (it turns out) the same predictions for any measurements performed with any kind of physical apparatus whatsoever. . . . they disagree as to whether that measured distance is the "real" distance, but such disagreement is a matter of philosophy, not physics. . . . Which viewpoint tells the "real truth" is irrelevant.

"There are two points I'd like to make here. First, Thorne is writing only with regard to gravity and general relativity effects. Lorentz showed that a 'flat space-time' interpretation of velocity effects (special relativity's domain) is also consistent with SRT's direct measurement predictions. Second, anyone who thinks that having two completely different explanations, mutually exclusive (but nearly identical in experimental predictions) should be of no interest to the theoretical physicist has to have a screw loose."

"Just a minute," David intervened, "Thorne didn't say 'nearly identical in experimental predictions', he said identical."

"I know what Thorne said," I replied. "But, clearly, he can't possibly know that they're identical in every respect. His lack of interest in exploring non-trivial implications is explicitly stated, emphasized even. In truth, the disparaged implications are sometimes more revealing than direct measurement.

"But even if they appeared to be identical, it seems profoundly stupid to have no interest in whether one or the other of two mutually exclusive interpretations is the right one. The only possible excuse for such an attitude is that one has 'swallowed the camel'. In any case, the two interpretations are not identical, but differ in very significant ways."

"What do you mean? Where's Thorne wrong? Give me an example of how they differ."

"The one interpretation (flat space-time, in Thorne's terminology) is essentially a subset of the MLET interpretation. And MLET, even in the form of its incomplete predecessor (Lorentz's ether based interpretation) clearly predicted the Sagnac effect — Sagnac expected the experimental result. SRT could never have predicted such a result and, in fact, in spite of some very clever rationalizations current theory still can't accommodate the results very well. Similarly, one interpretation clearly tells us that gravitational waves of the form predicted by conventional relativity theory will never be found — MLET clearly predicts that the LIGO experiments will not detect gravity waves. *Which viewpoint tells the 'real truth' is irrelevant?* What nonsense.

"Thorne is dead wrong. Direct measurement may not directly detect the differences, but non-trivial implications do differ significantly, and do have a real bearing on ability to predict. And I think even you will agree that ability to predict is a rather important characteristic of good theory."

David laughed. "I had hoped you might ease my frustration with having to deal with Steve. You haven't helped at all — if anything, you're even worse. I think I'd better call it a day."

On Good Theory
[Steven Mitchell and David Rhodes]
12 May, 1999

→ Have you given much thought to what we discussed yesterday?

♣ Frankly, not much. I think MLET is nonsense, and I think that I think that it's nonsense because it *is* nonsense. The physics community is much more open to reasonable change than you imply.

→ Give it a rest. I never claimed that the community isn't open to *reasonable* change. I only pointed out that truly iconoclastic change can never be considered reasonable from the consensus perspective. Openness to iconoclastic change is still an illusion.

And I don't fault this lack of openness to iconoclastic change. What I do find unacceptable, though, is the claim that if one is right, if one has *the goods*, the iconoclastic nature of the concepts would in no way preclude their enthusiastic acceptance by the elite community — that's deceptive nonsense.

♣ Even if I were to grant that, so what?

→ This kind of deception can be very damaging. It tends to mute strong objection by falsely assuring the skeptic that, if he's right, others will surely recognize the need for change and take up the banner for truth. It gives the impression that truth will out, always, and in the short term. I do believe that truth will out — eventually. But even with forceful advocates, we should expect that iconoclastic change may take decades — and without strong advocates, many decades. And no matter the strength of the evidence.

As you've implied earlier, our best hope of real impact seems to be within the interested lay community. But if they're convinced that anyone with the goods would be enthusiastically and immediately embraced by the elite community, they'll be much less likely to recognize very good science for what it really is. So it's important to us that this supposed openness to iconoclastic claims is recognized as the illusion that it is.

♣ I think you're wrong, it's no illusion — given 'the goods' rapid change is possible.

→ The null LIGO results may test your argument — and I hope you're right. But the impact of LIGO may very well depend on the ingenuity of the accommodation crowd in showing that the null results are consistent with current theory after all.

♣ Now you're being obnoxious. Null results would be catastrophic for current theory.

→ Am I? Would they?

A few years ago I heard of an article (I think it was in Physical Review Letters) that claimed that the Sagnac effect proved that SRT was correct. It was rather astounding to me that it got through any kind of peer review. Since SRT could never have predicted it, and since Sagnac's ether-based explanation was much more straight-forward, the only possible basis for claiming that the effect proved SRT correct would seem to be as follows: "Since we know SRT is right, any imaginable physical phenomena must, ultimately, prove that it's right — therefore, the Sagnac effect, being real, must prove that SRT is right."

♣ That's nonsense.

→ I think so. But look at it from the author's perspective — when you're absolutely sure that something's right, you do know, absolutely, that experimental data must agree with it. So extreme ingenuity is clearly justified when it's needed in order to accommodate problem data.

So I'm very sure that there'll be a great deal of effort expended to accommodate the null LIGO results; and whether or not they're successful will depend more on the strength of the prevailing consensus than on the reasonableness of the accommodation.

But enough of that.

Let's all play the skeptic. Can we agree on what we'd like to see in good theory? What do we see in theory that makes us think it's a good theory? I've said that MLET is superior to current theory, but why do I think that? You think SRT is superior. What is it about SRT that makes you think it's the better theory?

To the consensus community superior theory is rigorously limited to (first of all) full agreement with SRT's absolute symmetry. To get anywhere we must, at least tentatively, set that conviction aside. Having done that, I'd like to generalize from a true skeptic's perspective and ask, "What are the required characteristics of any superior theory?"

♣ That sort of question has been around for centuries. Why do you think you can add anything meaningful?

→ Let's try.

Let's first look again at the crucial differences between SRT and MLET.

Both SRT and MLET require an apparent symmetry with respect to any reference frame, regardless of velocity and regardless of the choice of frame. Einstein required that this symmetry be an absolute symmetry — the ultimate and complete reality. MLET requires the same apparent symmetry, but explains it as the reflection of a real, underlying asymmetry.

Einstein acknowledged that Lorentz, with his underlying asymmetry, had succeeded in showing that measured values (such as the Michelson-Morley results) were not inconsistent with an ether at rest. But, as we've pointed out in other contexts, in a speech in 1920 he[29] said of this underlying asymmetry:

> For the theoretician such an asymmetry in the theoretical structure, with no corresponding asymmetry in the system of experience, is intolerable.

In other words, if we can't experience it, if we can't measure it, then we can't tolerate it, we must reject it. Further explanation is not allowed. This is the positivist's perspective.

Max Born[30], in a similar manner, decries the introduction of a concept that "was introduced to explain." And continues:

> Sound epistemological criticism refuses to accept such made-to-order hypothesis.

These views remain the fundamental objection to the Lorentz interpretation of velocity effects. I consider these views arbitrary, based on no known objective truth, and not worthy candidates for required characteristics of good theory.

♣ You think that ad hoc concepts, introduced just to explain, should be tolerated?

→ Ad hoc? Who decides what that means? If a concept demonstrably improves understanding —

♣ Improves understanding? And who decides what that

means?

→ Well, let's revert to something we all understand. If the explanation makes unexpected new predictions, predictions that can be measured, and those predictions are subsequently born out by experiment, we have some reason to think it improves understanding. Right?

♣ Ability to predict is a positive, sure.

→ Good. With all due respect to Born, we like concepts that are introduced to explain, provided only that such explanation exhibits certain well defined characteristics — one such characteristic is consistency with respect to experimental data, the power to predict results that can be measured. So we both like consistency.

I take the position that it's possible for us to specify other pretty general fundamental preferences, characteristics we'd expect to find, must insist on finding, in any superior scientific theory. If we can agree on what characteristics we prefer, we may find a basis for eventual agreement on the characteristics of superior theory.

♣ I'd still agree with Davies' claim that, if relativity were wrong, our detailed understanding of subatomic physics would collapse. You'd have to address almost the whole of physics.

→ Remember we're skeptics. We're setting aside specific theories until we agree on what good theory is. But your remark does suggest another characteristic I'd consider important to good theory. An acceptable theory must address 'the whole of physics,' implying convergence — preferably universal convergence. A superior theory should have no arbitrary domain boundaries. We like convergence.

Let's look for other attributes.

Again, we like explanation. But unlike the positivist we don't pretend to think that direct measurement rigidly bounds good explanation. We do, however, insist on consistency and convergence — and more.

♣ Playing word games surely won't add much to ability to recognize good theory.

→ Humor me. Some characteristics I like to see in theory. Some characteristics you surely like. I think our preferences are likely very similar. So let me limit this morning's discussion to my preferences. I say preferences, because in the real world theories are almost never completely satisfactory.

The aim is to evaluate the quality (likely truth, plausibility) of explanation, and ultimately examine theory for excellence. I'll do that by examining how well my preferences are met. The better theory, the superior theory, is, to me then, the one that best satisfies my preferences.

The following characteristics, the following preferences, are my seven Cs of common sense.

♣ If I remember right, someone has defined common sense as "those prejudices we develop before our nineteenth birthday," or words to that effect.

→ Duly noted. But let's proceed on the assumption that Einstein isn't the only person capable of independent thought — I like informed common sense. OK?

♣ Let's see where you can take this.

→ I'm talking about deliberate, considered, freely acknowledged biases — preferences, guidelines, with respect to the nature of judgement itself. My common-sense preferences, are as follows:

- **Consensus** (agreement among those presumed to be most knowledgeable)

- **Consistency** (harmony of parts to a whole, emphasizing absence of contradiction)
- **Coherence** (systematic or logical connection, emphasizing strength and scope of connection)
- **Convergence** (to come together and unite in a common focus)
- **Clarity** (general eradication of confusion, understandable)
- **Concreteness** (specific, particular, unambiguous with respect to mathematical relationships; real, tangible, with respect to experience and experiment.)
- **Charm** (a compelling attractiveness, a trait that fascinates, allures or delights)

♣ Not much to object to. But rather trite don't you think?

→ Trite? No. Let's look at each in more detail.

Consensus. To some this may seem a strange bias. But, it isn't at all. Consensus among those we think have appropriate special knowledge is the single most common characteristic of what we normally consider common-sense belief. And no matter how much one may diverge from general consensus viewpoints, one nevertheless starts from some generally acknowledged base — at some level, consensus concepts always provide the essential context for sensible inquiry.

If one has no special interest and sees no reason for concern, a perception of strong consensus is convincing—the consensus view *is,* understandably, usually the common-sense view. There are many good reasons why this should be so. I have no special knowledge of poisons, yet I avoid them because people who should know tell me they're harmful. Most of our decisions in life involve a very high degree of dependence on what we presume to be true on the basis of what we believe with regard to a consensus of experts. Science, as it is generally taught (especially in the lower grades) is, and is intended to be, an introduction to consensus concepts. And that's as it should be. Significant consensus among experts can never be taken lightly.

♣ You like consensus, yet you clearly recognize that, with MLET, you're going against perhaps the strongest consensus ever found in any modern science. How do you reconcile this?

→ How do we reconcile going against consensus with an acknowledged strong preference *for* consensus? — an important question. In general, when, how and why might one justify going against firm consensus?

History warns us that even the firmest consensus should not generally, in itself, be considered sufficient to establish a concept as good theory. Consensus is a communal aspect of belief, and as such is subject to sociological forces — not all of which are always beneficial to good science. Interestingly, it's usually the very lack of consensus, or lack of confidence in a supposed consensus, that triggers significant inquiry — and, almost by definition, there can be no significant scientific advance without a reaching beyond consensus concepts.

There clearly have been times when strong consensus has ultimately been proven wrong. It would be difficult to justify any assumption that it'll never happen again. So we shouldn't completely rule out all challenge to consensus. But, when is it right to challenge? How can it be done safely?

The answer seems to lie in our more fundamental preferences. While consensus can't ever be taken lightly, other, more fundamental preferences, always allow us to reach out toward the unknown (or to re-examine the claims of even the firmest consensus) with some confidence.

So let's move on to the more fundamental preferences:

Consistency. By consistency we mean the harmony between parts and with the whole. Consistency emphasizes the absence of contradiction. Theory must be consistent with the experimental data — must be capable of predicting the measurable results of experiments not yet made.

But when new data seems to contradict old ideas (weakens theoretical consistency) and so triggers a challenge to a prior consensus, how we resolve the conflict will likely involve other preferences.

Coherence. By coherence we mean systematic or logical connection. Given consistency (a general absence of contradiction), we go beyond that and desire the additional strength of logical connection — a relationship, a relevance, of one part to another, of the new to the old. After all, various concepts can be consistent — involve no contradiction — and be virtually unrelated. But we generally assume that our universe is, in some sense, a single coherent entity. New knowledge must relate in some definite way to older knowledge. A new concept that has no relationship, or only a very weak relationship, to existing knowledge must, in general, be suspect as a bit too ad hoc to be enthusiastically embraced.

Convergence. By convergence we mean coming together and uniting in a common focus. It is, of course, very closely related to coherence, but here we prefer a common focus of a universal coherence. We find concepts more convincing when they're not only consistent with and strongly related to knowns, but also contribute to the understanding of underlying or overarching truths. The most convincing concept doesn't merely contribute to understanding within a separate, distinct domain with some arbitrary boundary, but is much more convincing when it contributes to a seamless logical unity at other levels. We prefer not to have to introduce arbitrary new fundamental, disparate principles (and consequent disparate domains) to accommodate a new concept, especially when such new principles seem incompatible with (or even just unrelated to) the ultimate focus of other domains of knowledge, of other convergent disciplines.

Einstein[31] once stated that,

> It is often, perhaps even always, possible to adhere to a general theoretical foundation by securing the adaptation of the theory to the facts by means of artificial additional assumptions.

A good observation. It's important to recognize the danger. We must have acceptable convergence toward a single basic understanding. The problem lies in differentiating between good explanation of newly revealed truth and merely an "artificial assumption" to accommodate that truth within the existing theoretical base. A useful question is: "Does the explanation give us improved universal consistency, coherence, and convergence?" If it does we tend to embrace it. If it does not, or if (worse yet) it further compromises existing desirable characteristics, we either reject it as an unacceptable 'artificial assumption' merely useful for accommodation, or at least hold it in abeyance until we learn more.

In general, given a good general theoretical foundation we should expect that new facts will strengthen overall conformance to our preferences. So, in keeping with Einstein's warning, if we cannot achieve adaptation of new facts to the existing general theoretical foundation without compromising our fundamental preferences, we must eventually question the quality of that very same foundation.

Clarity. By clarity we mean without undue complexity and with a minimum of ambiguity. Some might prefer simplicity rather than clarity but historically that seems to be dangerously misleading. All too often, when simplification has been the goal, insufficient attention has been given to broad implications. True clarity simply cannot exist apart from overarching consistency, coherence and convergence. (Occam's razor must be used with care.)

Clarity abhors all undue complexity, but especially complexity arising from overzealous prior simplification. History warns that apparent simplification may result in gross distortion of the proper interpretation of new knowledge — a progressive muddying of the understanding of underlying or overarching concepts — with the sure consequence that increased knowledge ultimately leads to increased confusion. So clarity is desired even where complexity is required. And it must be recognized that, because apparent simplicity is so seductive, oversimplification on some point, once generally accepted, can be extremely difficult to overthrow.

Concreteness. We're splitting the preference for the concrete into two parts because we have two necessary aspects to satisfy before we'll fully concede that we have a concrete concept. Concreteness may often be the final step in the achievement of general consensus, and so we might not expect a particularly high degree of concreteness in the early stages of a search for new knowledge — it is, nevertheless, a significant goal.

Mathematical Concreteness. Concreteness, in this sense, means specific, particular, unambiguous and rigorous with respect to the necessary mathematics. Common sense can't tolerate intractable mathematical contradictions or dubious mathematical procedures—mathematical integrity cannot be compromised.

But mathematics alone is not enough.

Physical Concreteness. Physical reality is conceptually both less rigorous and more restrictive than pure mathematics. As Richard Feynman[32] observed: ". . . many physical pictures can give the same equations." So, although we may be happy with a mathematical description, no matter how satisfying our mathematics may be, a certain ambiguity remains until our mathematics is mated with a unique and visualizable, concrete understanding of a physical reality. Mathematical integrity can't be compromised; on the other hand our universe is a single, tangible realization of only one of possibly very many imaginable mathematically consistent universes.

♣ Let's stop here just a second. You realize, I'm sure, that modern physics doesn't allow a "unique and visualizable, concrete physical reality." This sort of physical interpretation is soundly dismissed as naive visualization by any knowledgeable physicist.

→ Rather than argue that point here, let me just say that where visualization is possible, it's to be preferred. By that I mean that, given two explanations, one which provides an intuitively understandable (visualizable) physical reality, and another which is no better or no worse in any measurable way, I would prefer the intuitively understandable option. It remains a preference.

I said I didn't want to argue the point here, and I don't. But I should make one thing clear: I believe that the naive disallowance of 'naive visualization' may well be the most damaging aspect of Einstein's perfect symmetry. In discarding this discipline, physics has largely given up acceptable levels of consistency, given up any hope of coherence, and hopelessly compromised convergence and clarity. If, by some miracle, current theory with its rejection of visualization, came close to MLET in regard to the other preferences, I might consider accepting the common consensus with regard to visualization — but it doesn't and can't.

♣ You certainly have little respect for current thinking. I'll admit that I'm getting mildly interested in seeing where you can go with this.

→ That's a start.

♣ I didn't mean that as a positive. I think you're setting yourself up for a big fall.

→ I understood that. Still, interest, for whatever reason, is progress.

For the sake of continued argument, let me make a proposal. If, at the end of a month, you can convince me that current theory provides a more consistent, coherent, convergent, clear and mathematically concrete explanation of our knowledge base than MLET does, then I'll concede that current theory is better than MLET, in spite of current theory's lack of physical concreteness.

On the other hand, I'd expect that if I can convince you that MLET provides a more consistent, coherent, convergent, clear and mathematically concrete explanation of our knowledge base than current theory does, then you must concede that MLET is superior to current theory, even though MLET does provide a sensible, visualizable, concrete physical reality. All I'm asking is that, *all else being equal*, you never treat a sensible, visualizable, concrete physical reality as a negative.

♣ I can't imagine it ever coming to that, but I see no obvious reason to object.

→ Then let me repeat the point — all else being equal, a concept should not be discarded simply because it provides a sensible, visualizable, concrete physical reality. For our purposes, I'll only insist that physical concreteness be treated as a neutral characteristic.

But let's go on.

Charm. When a concept possesses charm we recognize a compelling attractiveness, a trait that fascinates, that delights. Charm encompasses beauty but has an appropriate negative connotation as well. We often think of charm as being somewhat superficial, and it sometimes is. Charm is never, can never be, completely objective — charm alone is never enough, and should be held suspect when other preferences are only marginally satisfied. While charm is so compelling that it may easily be given undo weight, all else being equal, common sense will not arbitrarily reject its allure — truth is, after all, usually quite attractive.

In Summary. The seven Cs are common-sense preferences. If these preferences are valid tools of inquiry then we should expect that knowledge builds on knowledge. To the extent that new concepts provide improved conformance to these strong preferences, we should expect that, as these new concepts are validated, unavoidable implications may work both up and down the knowledge hierarchy, revealing unexpected truths in areas previously thought unrelated or only incidentally related. In holding to the seven Cs, we would expect a rather high level of what is sometimes thought of as serendipity.

This aspect of knowledge development is described in one of Feynman's observations (as recorded by James Gleick). Gleick[33] said of Feynman:

> He had a set of practical tests, heuristics, that he applied when reaching a judgment about a new idea in physics: . . . for example, did it explain something unrelated to the original problem? He would challenge a young theorist: *What can you explain that you didn't set out to explain?*

This is the natural serendipity of the common-sense preferences.

Again, common sense would prefer that we progress toward a consensus that we have a consistent, coherent, convergent, clear, concrete and charming concept. In the real world of course, we must expect that few significant new concepts (at least when first proposed) will exhibit acceptable strength in all areas — and the resolution of conflict with respect to these strongly-held biases may not be a trivial undertaking. We are usually faced with both strengths and weaknesses, and giving proper weight to apparent anomalies may be a slow and tedious process. But fundamental conflict must ultimately be resolved — all preferences should be suitably satisfied — before we allow a confident claim of true understanding.

In comparing theories I'd add an eighth C. I'd add completeness. If two theories equally satisfy the seven Cs, which reaches the farthest? Which leaves the fewest unanswered questions?

- ♣ Aside from your physically concrete, I find your preferences appealing, generally desirable characteristics. But I'd be careful to note that we can't dictate to Nature. As Richard Feynman[34] writes:

  > I'm going to describe to you how Nature is — and if you don't like it, that's going to get in the way of your understanding it. It's a problem that physicists have learned to deal with: They've learned to realize that whether they like a theory or they don't like a theory is *not* the essential question. Rather, it is whether or not the theory gives predictions that agree with experiment. It is not a question of whether a theory is philosophically delightful, or easy to understand, or perfectly reasonable from the point of view of common sense. . . . So I hope you can accept Nature as She is — absurd.

  So it remains to be seen whether, or to what extent, your preferences can be legitimately and consistently applied to particular concepts. I have some reservations in that regard.

- → It's perfectly understandable that any intelligent person holding to Einstein's perfect symmetry must find Nature absurd. And if one knows that Nature is absurd, then one also knows that any intuitively understandable explanation of absurd phenomena is clearly absurd. So questioning current theory simply because it is absurd, is absurd — and the skeptic is left with little room for rational objection.

- ♣ That's ridiculous.

- → You're telling me. But that's the current situation. Although MLET provides not just equal, but superior ability to accurately predict, it's very reasonableness, the fact that it's intuitively understandable, can, from the SRT perspective, sensibly be perceived as proof that it's untenable.

Feynman's surety in the absurdity of Nature can only come from an unquestioning belief that no better explanation, no explanation with equal or better ability to predict, is conceivable. In that, he's wrong — MLET provides the superior interpretation. To us, it is absurd to think that, given two theories with equal ability to 'give predictions that agree with experiment' one should prefer the one that presents Nature as absurd over the one that presents our common-sense preferences as inherent characteristics of Nature Herself.

♣ As I said before, aside from your physically concrete, I find your preferences generally appealing. They are things we like to see — but as Feynman warned, we'll get nowhere if we can't accept Nature as She is.

→ Fine. For you, they are *only* preferences — but *real* preferences nonetheless. If you merely hold to them as nothing more than that, we think you must eventually come to prefer MLET's underlying asymmetry to Einstein's absolute symmetry.

♣ Again, I like your preferences but I still agree with Feynman: "It is not a question of whether a theory is philosophically delightful, or easy to understand, or perfectly reasonable from the point of view of common sense. Rather, it is whether or not the theory gives predictions that agree with experiment."

→ Consistency clearly requires predictions that agree with experiment — that's a given. But doesn't it bother you at all that current theory is so lacking in coherence and convergence? We only argue that it is entirely possible that more than one interpretation may make essentially the same predictions — and in such a case, superior coherence and convergence should be telling.

So let me just summarize my objective for this session.

We set out to distinguish between good judgement and mere ingenuity in accommodating new facts within the constraints of a given theoretical base. We're looking for a disciplined means to judge the acceptability of explanation, the excellence of proposed theory. If we have succeeded we then have a means to compare theories, setting aside any initial biases with regard to which (if any) is true. I call it resorting to common sense.

In the most formal sense permitted by the nature of ourselves and our inability to directly perceive reality, I would define this common-sense approach as a system of applied preferences — carefully considered biases. Some things, things fundamentally compatible with the human mind, we have come to prefer, to like, to respect — some things repel us. Consistency. Coherence. Convergence. Clarity. Concreteness. These are concepts we generally like, concepts we call sensible. And to possess charm is, by definition, to be appealing. And where personal knowledge is lacking we look for consensus. These things we prefer.

So to the extent that sensible preferences are compromised, we will tend to reject (or only reluctantly and tentatively accept) a proposed theory. If the required compromise is extreme, we will find claims unacceptable and reject them. Conversely, no matter how satisfying a theory may be, if we find that some alternate theory better satisfies our preferences we will prefer it. This, to me, is the essence of common-sense.

♣ I suspect you've provided a rather solid basis for the rejection of your claims. I'm sure that as long as you prefer superior ability to predict, superior consistency, there's no way you'll do better than current theory with respect to your preferences.

→ We'll find out. I'll see you back here tomorrow.

— *** —

David came into my office, smiled and asked, "More coffee?" When I nodded he filled both cups from the still nearly full carafe.

"So what now?" I asked.

"Consensus, consistency, coherence, convergence, clarity, concreteness, and charm." David mused, "I can live with that. I'd bet MLET can't. But, enough for now; let's finish our coffee and go for a sail."

The day and the lake were beautiful. We didn't return to the dock until well after lunchtime.

What Can We Ask of Theory?
[Steven Mitchell and David Rhodes]
13 May, 1999

→ We've covered quite a lot in these first hours.

Yesterday, we discussed what I feel are the key characteristics of good theory — our common-sense preferences. The day before we discussed the limits of openness to change.

♣ You pointed out a couple days ago that I didn't know much about MLET. That's still true, so when are we going to cover the actual theory? We seem to be skirting around the edges.

→ Well, we (Julia, Jim and I) went 'round and 'round on how we should use these thirty days, and we all agreed that there was more danger in presenting the theory too early than in presenting it too late.

♣ Why's that? If you don't mind my asking . . .

→ Not at all. Just the fundamental principles of brainwashing.

Seriously, we've bought thirty hours of your time. By limiting ourselves to an hour a day we can get away with considerable repetition of key points. And whether you like it or not, you'll probably actually think about what we're saying a great deal more than the one hour. So thirty times over thirty days is much better from our standpoint than one time over three or four days.

As we see it, our biggest problem has little to do with the reasonableness, the merits, the strength of the goods, if you will, of MLET. Given a chance it'll stand on its own merits. The much more serious problem is how do we overcome extreme prejudice. Without some slight recognition that SRT is not absolute fact, we'll have no chance. To that end we decided not to talk specifics until we detect some semblance of openness, a willingness to listen to our claims.

Or until about the tenth day, whichever comes first.

♣ If by willingness to listen you mean an inclination to take your claims seriously, I'm sure the tenth day will come first.

→ Whatever. I do want to spend a few more days with you before I present anything more than general claims.

♣ So where do we go from here?

→ Let's continue our attempt to give you a better sense of where we're coming from. If you don't mind, I'd like to spend this morning with some additional general observations with regard to current theory and the common-sense preferences we discussed yesterday.

♣ General observations? What does that mean?

→ More of the same, I guess. To review what we've already covered from a slightly different perspective. MLET makes some specific predictions: among these, no gravitational waves of the expected form. Of course that's not all it does. It conflicts with current theory. But the conflict is with theory, not with the data or the supporting mathematics. MLET is in no way destructive of clear understanding. In every area — special relativity, quantum theory, general relativity, whatever — MLET significantly enhances understanding of the experimental data with almost no change in the essential mathematics.

We'll show that these benefits arise from the simple reinterpretation of velocity effects — rejection of the special of SRT in favor of the universal of MLET. Philosophically, the real difference between the current SRT constrained worldview and the MLET worldview, is that MLET unequivocally rejects positivism as total nonsense, while SRT has no foundation apart from it.

♣ You're aware that Stephen Hawking admits to taking the positivist's position with regard to the limits of theory?

→ I'm aware of that. That's his problem — but certainly not his alone. I would argue that the only possible grounds for rejecting Lorentz's better base in favor of SRT is that Lorentz went beyond the positivistic constraints to provide understanding — so no matter what one claims to the contrary, any theory that embraces SRT's absolute constraints, is a theory based on positivism. Take away positivism and not only do serious questions arise with regard to SRT concepts, but, with the questions now allowed, the better theory becomes obvious. Hawking, I think, might be one of the few to acknowledge that current theory does disallow questioning in precisely the manner of strict positivism. Hawking[35] writes:

> I take the positivists viewpoint that a physical theory is just a mathematical model and that it is meaningless to ask whether it corresponds to reality. All that one can ask is that its predictions should be in agreement with observation.

As you can imagine we reject this view as nonsense. Although Hawking's position seems somewhat extreme, even to some physicists, his position seems to me to be the only position open to an honest advocate of SRT. SRT could never survive any conceivable alternative to the positivist's disallowance of the telling questions.

- ♣ Philosophically, I'd have to say I don't give a damn. I'm a scientist, not a philosopher. I think you're wrong, but I don't much care for this kind of metaphysics.
- → Yet, it was clearly positivism that led Einstein to his absolute —
- ♣ Einstein was the one who did away with absolutes. I find your persistent use of that term with respect to Einstein's work offensive.
- → It was positivism that lead Einstein to reject any underlying asymmetry and insist on *absolute* symmetry. His absolute is a mystical, magical, ghostly absolute — but no less an absolute. And, unlike the absolutes he rejected, it wasn't experiment, it wasn't observation, but an intuitive leap, a subjective preference, that led to Einstein's position.
- ♣ I have to listen to this?
- → Yeah, you do. So bear with me.

Leon Lederman[36], in a statement remarkably close to Hawking's, (clearly reflecting the undeniable positivistic base from which modern physics theory springs) comes close to the consensus opinion in his book, *The God Particle* when he writes:

> All we can ask of a theory is to predict the results of events that can be measured. This sounds like an obvious point, but forgetting it leads to the so-called paradoxes that popular writers without culture are fond of exploiting.

I love his 'forgetting it.' Without culture, one could never believe it — the culturally deprived have nothing to forget. In our view, this "all we can ask" nonsense is the very root of the intellectual bankruptcy of special relativity — positivism's pathetic legacy. 'All we can ask' may well preclude serious consideration of apparent paradox, but 'all we can ask" is itself, simply unacceptable.

♣ Don't mind me. It just struck me as very amusing that you would presume to tell Lederman what is or isn't acceptable. Do you know who he is?

→ He's a great physicist, a truly outstanding experimentalist, a great teacher and a great writer. And I'm not being the least bit facetious when I say that — I loved the book. But we'll get nowhere if we can't express our convictions frankly. Nothing personal — all we're saying is that the statement is overly simplistic — even most positivists, when pressed, would generally acknowledge that there is always a need for coherence, convergence, clarity and mathematical integrity. This need can't reasonably be casually dismissed as more than can be asked.

♣ That's insulting. You can't honestly presume any intent to deny the importance of any of those attributes.

→ Use your head — the 'all we can ask' claim only need be made, only makes sense, if one perceives a need to justify acceptance of the obvious absence of characteristics otherwise thought essential. That absence must somehow be made tolerable.

How's it done? Are essential attributes missing? — too bad. It's tough, but, sorry, 'all we can ask,' really means exactly that — you can't ask Nature for what Nature elects not to provide. That's the way it is. Live with it.

I say that's unacceptable.

♣ No one goes so far as to dismiss essential attributes.

→ They really do. Of course, as we just pointed out, they argue that 'essential' can't exceed what direct measurement provides — they simply can't allow

Nature to exceed the bounds of man's ability to directly measure

Just look at the matter of emphasis, of focus. It's true enough that the requirement for consistency is never satisfactorily met if empirical data is in conflict with theoretical values; so, yes, theory must be able to 'predict the results of events that can be measured.' But the 'all we can ask' focus does, and I think is intended to, minimize the impact of obvious weaknesses. The real lack of coherence and convergence that is so obvious in the intractable domain boundaries of current theory shouldn't be so blithely dismissed — especially given the glaring lack of empirical support for SRT's claims.

We do ask more.

If one can encompass almost all of current knowledge in a single consistent, coherent, convergent, clear and concrete whole, then one has the best possible base for making predictions as to the results of experiments. And it's the very lack of so many of these characteristics that makes current theory so impotent with regard to prediction much beyond experimental frontiers. Trying to go forward without first understanding what is already known can be a very frustrating process.

- ♣ So you can make better predictions than current theory can?
- → Sure.
- ♣ And, as you've repeatedly pointed out, only MLET predicts —
- → Yes, we've repeatedly mentioned one solid prediction — the gravitational wave experiments will find nothing. One interesting thing about this prediction is that

MLET was developed with no thought of trying to predict the nature of gravitational waves. It simply turned out that the better understanding of the nature of the ether, of matter and of electromagnetic waves, clearly required the new prediction.

♣ How're you going to feel when so much of your fifty million dollars goes down the drain? It's life or death for MLET.

→ We don't see it that way at all.

♣ What other way is there? Verified detection of gravitational waves would kill you.

→ But that won't happen. The LIGO results are not life or death for MLET — they're validation.

♣ That level of confidence is ridiculous.

→ Well, the evidence is very clear and very strong. I don't feel any anxiety in anticipating the results. But how will you feel about the many hundreds of millions of dollars spent in the hope of detection?

♣ It will prove to be money well spent.

→ I'm glad they're spending it, but they won't like the results. But we've been over all this several times already, so let's get back to our discussion of positivism and SRT.

As I've claimed, take away SRT's positivistic base and you have no SRT — it stands or falls with positivism. And positivism, the 'all we can ask' mentality, provides no basis for meaningful interpretation. But what really throws one for a loop is Lederman's, "This sounds like an obvious point." It really brings home how culturally deprived we MLET advocates are — to me the obvious point is that 'all we can ask' is a pathetic cop-out. A dismissal of confounding problems by simply denying their right to

exist.
- ♣ You just can't accept the nonintuitive implications of SRT. You can't face the difficult truth. Admit it and let's all go home.
- → Difficult truth? The *sad* truth is that a physics with SRT at its roots is insanely deficient. Where's the consistency? Where's the cohesiveness? Where's the convergence? Where's the clarity? Where's the mathematical integrity? And you pretend I'm the one unable to face implications.

    With all due respect to Lederman, given two interpretations, each with only equal ability to predict results of events which can be measured, we would almost universally prefer the one that gives us the best convergence, the strongest coherence, the most clarity and the greatest degree of consistency — we ask more. Again, we see Lederman's 'all we can ask' claim as a simple assertion that, in spite of its obvious deficiencies, the current worldview is absolutely the best that one can hope for — again, live with it.
- ♣ If better were possible we would've discovered it by now.
- → Of course, so go back to sleep. Since you know you have the best possible, we'd better not suggest that it's not. Deficiencies? That's just the way nature is.
- ♣ And you're modern flat-earthers. You've said that SRT has no foundation apart from positivism, yet there's simply no doubt that the experimental data supports SRT.
- → You're living in a dream world. Experimental data strongly favors the sensible alternative.
- ♣ I'm living in a dream world?
- → Somewhere, sometime, I think we may just get to you.

SRT is roughly consistent with much of our empirical base, but it violates every tenet of good science. It can't give us coherence. It destroys any hope of universal convergence. It's nonsense.

Look, positivism can't go beyond measured values. As Mach pointed out, direct measurement can tell us absolutely nothing about absolute space and universal time. Positivism, by extension, can tell us absolutely nothing about space and time. Yet Einstein presumed to tell us something very profound with respect to both space and time. Where did his knowledge come from? Certainly not from experimental data.

It's easy to see why some have called his insight an intuitive leap. I agree. It was an intuitive leap into an intellectual abyss.

♣ No matter how you slice it, we know Einstein was right — SRT works.

→ And MLET works better. If you were absolutely restricted (as positivism prefers) to direct measurement you would almost never see the difference between the two. But explanation, and so ability to predict, goes beyond such arbitrary constraints and clearly favors MLET. So don't give me that 'SRT works' bit.

Really, where did Einstein's knowledge come from? Not from direct experiment. Where then? It's knowledge by edict, based solely on an inability to accept good explanation because of some quaint notion of what is tolerable. To accept Einstein's interpretation as proven fact, one must first reject the better explanation.

♣ That's crazy.

→ You're right. In SRT modern physics embraces one crazy, contrived assumption.

- ♣ Don't twist my words.
- → Don't twist the reality. Theory can't even be properly thought of as theory if it refuses to go beyond the directly measurable. In SRT, Einstein did go beyond the silence of positivism (as all theorists, even avowed positivists, must) by arbitrarily selecting one dubious, unproven and unprovable, alternative. He declared that unavoidable ambiguity was certain knowledge and simply asserted that measured rate of change completely defines any possible understanding of time itself.
- ♣ What practical difference does any of this make?
- → What difference? He can't allow the crucial questions that we insist on asking. We assert that measured values tell us something about real phenomena, phenomena reflecting the real behavior of a physical reality, a reality that in turn demands an underlying asymmetry. He's forced to reject all of that because he can't tolerate the essential asymmetry. We require an ether. He can't allow it. We require consistency, coherence and convergence, he limits theory to ability to predict. He can't get past disparate domains, we find universal convergence. And on and on.

The positivist's position involves both a very severe restriction on the allowed interpretation of experimental results and a significant undermining of the reasonable demands of intellectual discipline and fundamental integrity. But even if one does ask no more of theory than that it make predictions that can be measured, even then, MLET is superior.

In SRT a contrived ingenuity supplanted good judgement. And so modern theory is forced to play fast and loose with almost every aspect of intellectual discipline. Very good experiment may keep allowed results reasonably accurate, but, brother! — what the current consensus does allow in the name of theory!

♣ I dislike positivism, but I find your characterization of current theory offensive and ridiculous. And I certainly don't accept your claim that SRT is not supported by direct experiment.

→ Whatever else you may get from us, I do think that at the end of the month you'll be forced to acknowledge that, at the very least, SRT is certainly no better supported by direct experiment than is the clear alternative. And, if you ever find it possible to entertain the reality of an underlying asymmetry then you'll find the clear alternative more comprehensive, more consistent, more cohesive, more convergent, more concrete —

♣ Give it a rest!

→ Given alternatives, why reject the best?

♣ Look — Einstein himself had rejected the positivist viewpoint by 1930, but he certainly didn't reject SRT.

→ No, he didn't reject SRT. I suspect that he never acknowledged, even to himself, that in rejecting positivism he had discarded all arguments favoring SRT, and by extension, favoring general relativity also. However, as he grew older he did come to realize that his fundamental concepts might be in serious trouble.

♣ Where'd you get that?

→ In his book, *Albert Einstein: Creator and Rebel* Banesh Hoffmann[38] writes of Einstein:

> To Solvine, who had written congratulating him on his seventieth birthday, he wrote in reply on 28 March 1949, saying in part: "You imagine that I look back on my life's work with calm satisfaction. But from nearby it looks quite different. There is

not a single concept of which I am convinced it will stand firm, . . ."

♣ The man was seventy years old and had wasted over two decades working on his unification theories. I'm sure he was understandably frustrated by his failures. You shouldn't take the old man's frustration too seriously.

→ On the other hand, maybe you shouldn't take the plain statement too lightly. Why do you think all of his efforts at unification failed so miserably? He may have finally conceded (to himself at least) that his absolute-symmetry interpretation might be the source of the intractability of his unification problems.

♣ Whatever the reason for Einstein's repeated failures, you can't make a case against SRT based on the failure of a supposed positivistic foundation.

→ Anyone could.

But, as I suggested earlier, even positivism didn't really support Einstein's claims well enough to be considered a true foundation. Positivism, on its own, doesn't really *support* anything — its contribution is never more than a supposed justification for the denial of explanation. Positivism's impotence really renders it mute with respect to either interpretation — both go beyond what direct measurement can tell us. But only positivism, by denying the acceptability of the better explanation, could make Einstein's claims tolerable.

In any case we can do better.

♣ But what about the successes? You can't deny the scope of the advance of physics based on SRT.

→ To the extent that SRT and MLET required the same measured values and made the same predictions, SRT did little harm. But once theory moved beyond simple velocity effects, SRT's constrictive weaknesses became troubling. Confusion has been its legacy from rather early in this century, and in the last few decades SRT has been a major negative. The realized advances would almost certainly have come at least as fast if theory had stayed with the Lorentz ether — and with very much less confusion. The more recent advances have come in spite of theoretical weakness, not from strong theory — almost entirely from outstanding research.

♣ That's unfair. Solid progress itself has made it more and more difficult to test advanced theory, so even where theory gives strong predictions, these predictions are very nearly untestable.

→ Your opinion. At this point I think real advances will only come once we better understand what we already know.

♣ What kind of doublespeak is that?

→ There is an immense knowledge base to draw on. But understanding is very weak.

♣ And your solution?

→ Seriously re-examine basic concepts. That hasn't been done for nearly a hundred years. It's time. You've had nearly a century to develop an understanding based on SRT. What do you give us? Why (with the notable exception of consistency, coherence, convergence, a rigorous mathematics and a physical reality) whatever is needed, modern physics can deliver. Need a new particle? Strings? Superstrings? Balloons? A marvelously non-vacuous vacuum? Modern physics, with its 'all we can ask' mentality, is so well equipped to accommodate surprising experimental results that it's hard to even imagine any experimental outcome that couldn't be rather easily accommodated.

This adaptability is not a sign of strength. We have a community of theoreticians standing by, anxious, ready and able, to provide an accommodation of any conceivable experimental result. And as long as we hold hard to 'all we can ask' they'll probably succeed.

♣ That's absurd.

→ Not at all. Just watch what happens over the next few years. As I said earlier, I'd bet that, while the null results of the LIGO type experiments may provide a temporary challenge to the consensus worldview, the community will find a way to accommodate those results within the constraints of current theory. Someone may even win a Nobel prize showing how those null results, when properly understood, once again prove SRT to be correct.

♣ You're coming across as an arrogant, insufferable dolt!

→ OK, let me back off a bit — but just a bit. The point is, this current ad hoc conglomeration of disparate domains, this confusing patchwork of almost independent and very nearly mutually exclusive theories called modern physics, is not the best that we can do. It's no wonder that physicists are plagued with so much bothersome noise, so much nonsense, from the lay public.

♣ I've said it before, I'll say it again; most of that nonsense comes from ignorance like yours — a perception that all of nature must, in the final analysis, be intuitively obvious. Nature isn't like that. It's just very hard for many people to grasp the true significance of relativity.

→ You're partly right. There are some rather fantastic displays of ignorance out there. But I would still argue that if current theory came even close to giving us a consistent, cohesive, convergent and mathematically rigorous understanding of reality, or even showed any promise of ever doing so, the incessant noise would diminish to an acceptable level. Most of the noise comes from the fact that almost any reasonably intelligent person can see that something is seriously wrong.

♣ So we're all idiots? Tell me, what makes someone like me, someone with a very good education and exceptional experience, such a poor judge of truth?

→ In some respects, your education is your problem.

♣ Then why didn't you pay some high school drop-out to listen to your nonsense?

→ Education is expected to, and so usually does, re-enforce the *settled* beliefs of the controlling culture. How could it be otherwise? So yes, if the consensus view is wrong, we should expect that education may contribute significant negatives. No surprise there.

Michael Levine[39] is at least partly right when he writes in his book, *Lessons at the Halfway Point*, "Some ideas are really so stupid that only intellectuals could believe them." He has a valid point. Over time, firmly established and repeatedly affirmed consensus can effectively eliminate reasonable skepticism. And in the case of a student of physics, if any skepticism with regard to SRT isn't sufficiently overcome by the time one has a bachelor's degree, the sensible student will turn to some other discipline.

♣ So you're claiming that my education has hindered my understanding of physics.

→ Inasmuch as we disagree with some of what you were taught (and so believe) yes, we do think that your education has been a hindrance to understanding. Your advanced studies have surely contributed to your inability to question SRT — and, from our standpoint, that's a big negative.

However, your considerable mathematical skill, your awareness of experimental data, of mathematical relationships, your acquired discipline, your obvious ability to learn, are all significant accomplishments — direct benefits of your educational experience.

I've emphasized the negatives of the current situation in order to highlight the room for improvement, but I don't intend to disparage individual accomplishments. Experimentalists, especially, have kept physics moving in spite of weak theory. So who knows? You certainly wouldn't be of much use to us without your hard-earned credibility within the culturally elite community.

- ♣ I'd be very surprised if my month here helped you in any way.
- → I initially opposed you're coming, but I now think there's no doubt that we'll benefit from your presence.

We'll gain some sense of whether (to what extent, and under what circumstances) any credible mainstream physicist, even a young one, might ever entertain the possibility that Einstein was wrong in any significant manner. We're quite interested in some answers in this area. Without your background you wouldn't be credible, and so we wouldn't care whether or not you could entertain doubt.

- ♣ So whether or not I ever question SRT you'll have learned something?
- → Yes. To me, that's what this's about.
- ♣ And if you fail with me you'll not bother other

physicists?

→ Price has made a major commitment in this area. We may change our tactics, but quit? Not very soon. You're the first but you won't be the last.

♣ As I said the other day, my concern is that you may cause serious confusion within the lay community.

→ I don't think we'll be taken seriously enough in the short term to trouble anyone. Our lack of culture will be easily recognized.

Seriously, regarding you . . . I may be overly skeptical but I think that, at your level, there's little chance that you could ever find us credible. I think Dingle was certainly right on one point — at a certain point the modern physicist simply loses any ability to question SRT.

♣ And regardless of any contrary evidence?

→ I wouldn't put it that way exactly. One simply becomes very sure of his knowledge and so would never take the time to seriously examine any contrary evidence. Amounts to the same thing, though, doesn't it? So, yes — regardless of any contrary evidence.

I really would like to be wrong with regard to your ability to entertain doubt. But a word of warning; finding us credible would be very close to professional suicide.

♣ You don't think my credibility would survive?

→ If you openly expressed serious skepticism with regard to SRT? — my opinion? — absolutely not! You might occasionally get away with presenting sympathetic views by carefully qualifying those views with statements such as: "We know this interpretation conflicts with SRT and so is known to be incorrect, but it does exhibit a surprising degree of agreement with experimental data." But you'd better not even do that sort of thing very often.

Seriously, watch out. What we want to learn from you is: "What can we do to you?"

♣ You flatter yourself if you're concerned about me ever thinking you credible. It's not gonna happen.

→ I think you're right. We'll find out. And the results will help us determine how much effort should be expended trying to directly influence consensus thought.

♣ Give it up. You don't have a prayer.

→ Price pays the bills, and he insists that we try. So I'll see you tomorrow.

—***—

With Mitchell gone, David turned toward the one-way mirror with his now familiar 'is this really happening to me' look on his face, and before even getting through my office door, said, "I'm stunned, perplexed, exasperated and absolutely amazed."

"That's progress," I said.

"Look at you people. You're all rank amateurs playing at science, presuming to challenge the best and the brightest, the leading experts in the most complex of sciences. It's laughable. What can you think you have to offer?"

"Money," I suggested with a laugh.

"It's not funny," David insisted. "Price is spending a fortune to confuse the lay public — that's what this ridiculous project really amounts to. You'll get nowhere with the physics community, but if you're half as clever as you seem you just might do serious damage by compromising funding for important experiments."

"Then don't you think it's important that you really understand every nuance, every possible implication, of what we're saying? Funding for very important efforts may depend on solid, credible refutation of our claims. Who better to do that than someone who really understands those claims?"

"You really are absolutely nuts," he replied. "You say that

as though you're certain I won't be able to discredit them."

"Yeah, I'm sure you can't," I said. "But seriously, if you're right, and since we're prepared to spend at least fifty million dollars to promote our viewpoint, don't you think it's important that you give disproving that viewpoint your best shot?"

"Julia, you're really getting to me. I thought Mitchell was bad, but you're even worse. You don't even take my concerns seriously. Can't you understand how much damage you people might do?"

"Let me be clear," I said. "We're spending thirty days of our time and more than thirty thousand dollars of our not unlimited funds on you. We take this very seriously. That said, we want most of all to get you to listen. If you're determined to refute our claims you *will* listen. So your determination to discredit us doesn't bother me in the least. I welcome it. If you succeed, it'll be because we deserve to be discredited. Do I take your concern seriously? Sure. The more concerned you are the better we like it. So if I can further motivate you with a little gratuitous needling, you can bet I'll try."

"You do have a point," David acknowledged. "If it weren't for the damage you might do, I wouldn't give you the time of day. Trouble is, I'll look like a fool if I take you seriously, even if only to discredit you."

"Oh, come on. You don't even need to acknowledge that you're aware of our existence unless we start having an impact that you don't like. If you ever sense a need for solid refutation of our claims, then, and only then, you step forward and become the hero," I suggested.

"Cute. But that doesn't change the fact that my current situation is ridiculous."

"How's that?" I asked.

"I have to listen for another — what? — twenty-seven days. Obligated to treat nonsense seriously for nearly four more weeks!"

"Isn't it disgusting what some people will do," I mused, "for a measly thirty-thousand dollars and an all expenses paid month-long vacation at Tahoe."

"Do you play tennis?"

"Sure," I replied.

"Can you meet me at the upper courts at ten o'clock?

I could. I did.

He was a much better player than I expected — obviously didn't spend *all* of his time on physics theory. It was a nice break — we played until lunchtime..

About Time
[Steven Mitchell and David Rhodes]
14 May, 1999

→ Let's briefly review where we are and where we hope to go.

Ultimately, we want to show that MLET is markedly superior to SRT — no hurry on that point. We first want to create room for doubt with respect to SRT. Just an ability to question.

♣ And so you dream on.

→ Let's continue.

Again, if you find weakness (with respect to consistency, cohesiveness, convergence and the other preferences) in our claims, fine — just don't think that disagreement with SRT is sufficient in itself to justify rejection.

♣ But of course it is.

→ You'd reject a concept that, when compared directly with SRT, demonstrates improved ability to predict things that can be measured, superior convergence, a stronger coherence, and improved consistency, simply because it disagrees with SRT?

♣ If you could show that it does all that? No. But that's clearly impossible.

→ But if we do the impossible?

♣ I'm not really into playing pointless games.

→ We'll never get anywhere if you turn off every time we suggest that disagreement with SRT is not, in itself, a fatal flaw. We do, we do, we do, reject SRT as the ultimate standard.

Can you get past that?

♣ It would take a lot more than you've shown me so far.

→ We'll show you a lot more. Slow and easy, repeating, repeating, repeating. And if you can never concede that SRT just might not be the final, absolute truth you currently think it is, we can at least make you suffer.

In the next two days we'll look briefly at two points of contention.

Today, the concept of time.

Tomorrow, the ether itself.

♣ Can you imagine what I'd have told Price he could do with his money if he'd given me a better idea of what I'd be listening to?

→ I think I can. But you're here so get over it.

About time. Mach claimed that the concept of a uniform passage of time throughout the universe had neither a practical nor a scientific value. Einstein went much farther. He claimed to know something very profound about time — he claimed that it was completely defined by measured rate of change. This declaration went well beyond what any possible experiment could ever show. In special relativity he treated absence of direct proof as direct proof of absence — a convenient fiction which allows the positivistically inclined to disallow explanation.

♣ Metaphysics, here we come! We've been over this before.

→ And we may very well return again, and again, and again. Look, I couldn't care less what you label the difference between SRT's and MLET's concepts of time. The difference is real and significant. Einstein (and so current theory) couldn't allow time to be independent of measured rate-of-change, and to the positivist it's all just metaphysics anyway, but so what? In even offering the alternative we are necessarily rejecting positivism, and without positivism to make explanation intolerable, the better explanation becomes acceptable; and once acceptable, much preferred.

♣ You're going way beyond what experiment can directly tell us.

→ Of course we are. So did Einstein. So does SRT. So does general relativity. So does anything properly labeled theory. We're going for explanation — explanation that gives us consistency, cohesion, convergence and all the rest. You can take your metaphysical charge and shove it. The better explanation is the better theory. Let's proceed on that basis. OK?

♣ So show me the better explanation.

→ Good. Back to time. Isaac Asimov[40], discussing Einstein's concept of time wrote:

> A clock in motion, he said, keeps time more slowly than a stationary one. In fact, all phenomena that change with time change more slowly when moving than when at rest, which is the same as saying that time itself is slowed.

The claim (that the slowing of the rate of change of all phenomena that change with time is equivalent to the slowing of time itself) cannot be the subject of experiment, so it's ridiculous to claim that the observed slowing proves that Einstein was right. No direct measurement, no possible experiment, can ever tell us that 'time' and physical 'rate of change' are identical concepts. Einstein claims that they are, but does Einstein's opinion really matter that much? Why should we take any such claim seriously? Why should anyone think that the real slowing of a clock, tells us anything about time itself?

With time independent of physical rate of change, we can explain why clocks slow and why rulers contract, why the measurements are what they are, and in doing so we get very strong theory; or, with Einstein, we can insist that time itself is nothing more than measured rate of change. With the Einstein option there can be no questions, there can be no explanation. What you see is what you get. Simple. But, oh man! — are you ever in for a wild ride!

♣ You can't deny that current theory, treating space-time as Einstein specified, when consistently and properly applied, gives the right answers.

→ That again! The right answers — everybody knows that good theory must give the right answers. And, yes, current theory can be tortured into yielding the right answers.

So who cares about our preferences?

Consistency? Nice but not possible across domains (so no longer essential).

Cohesion? Sure, we'd like it, but Nature insists that we accept Her as She is.

Convergence? The reality revealed by modern physics gives us disparate domains — can't you appreciate how liberating this can be to the accepting physicist?

Clarity? Well, Nature is absurd — but don't fight it, enjoy it.

Mathematical integrity? Well, almost. Anyway, those intractable infinities are probably just another verification of the delightful absurdity of Einstein's

Universe. Don't let it bother you — physicists have learned to live with it.

Sure, if it were possible, we'd love to have these things, but if we can't, well, life goes on. Who's naive enough to think our imagination is necessarily adequate to our task? At least we can rest in the sure knowledge that in SRT we have the best possible foundation; and since it's the best possible, it's gotta be good.

Maybe you like that kind of nonsense — we don't. Sure, we'll get the right answer if the mathematics is consistently and properly applied. Yes we will. But please don't ask what 'consistently' and 'properly' really mean in this context.

♣ What's that supposed to mean?

→ Unfortunately, in all too many cases, these terms mean little more than "We know how to manipulate the numbers to get the right answer — so just follow directions."

♣ Ridicule if you want, but Einstein's base has served us remarkably well for nearly a hundred years.

→ It's frightening to realize that a lot of otherwise very bright people actually believe that.

♣ Again, you may not like the theory, but it does work.

→ Yes it does. Feynman[41] in his book, *QED*, sorta summarizes modern success.
> The more you see how strangely Nature behaves, the harder it is to make a model that explains how even the simplest phenomena actually work. So theoretical physics has given up on that.

♣ That's the reality.

→ I don't like quitters. Who, and by what rights, dictates that we give up?

♣ Physics 'has given up on that' because it's been

repeatedly proven that visualizable physical models are utterly impossible.

→ It's been repeatedly proven that visualizable physical models are utterly impossible *within the bounds of Einstein's symmetry*. Yes it has. We know that's true.

♣ So why can't you accept it?

→ We do accept it as an unavoidable ramification of SRT — but that's SRT.

With the Lorentz asymmetry we have a much stronger physics *and* a visualizable physical reality, so it'd be silly to accept current theory's failure as anything more, or anything less, than proof that SRT is insanely deficient. Why would one choose to? We like visualizable physical models, so why should we throw them away just because someone tells us that we must? No experiment favors Einstein's absurdity.

♣ So if we give up Einstein's 'no preferred frame' everything's suddenly explained.

→ I'd never claim that the new perspective automatically, and without any effort on our part, provides intuitive understanding. What I do want to emphasize is that from SRT's perspective, as you've acknowledged, a visualizable physical model is known to be impossible. On the other hand, with MLET, our successes to this point, have convinced us that we have every reason to believe that a visualizable physical model of all phenomena, including all quantum effects, is achievable — giving up is, clearly, no longer a sensible option.

That's a very significant improvement.

♣ And exactly how do we get to this brave new world?

→ First, get rid of all traces of positivism.

♣ And where does one find these 'traces of positivism.'

→ In modern physics? Wherever you look. As Max Born[42] wrote:
> A concept refers to a physical reality only when there is something ascertainable by measurement corresponding to it in the world of phenomena. This is not the place to enter into a discussion on the philosophic concept of reality; it is at least certain that the criterion of reality just given corresponds fully with the way the word "reality" is used in the physical sciences. Every concept that does not satisfy it has gradually been eliminated from the structure of physics.

♣ That's not positivism. That's just rigorous definition of meaning.

→ That's positivism in its purest form. Born immediately follows the above with,
> We see at once that in this sense a "fixed spot" in Newton's absolute space has no (physical) reality.

And Steven Weinberg[43], in *Dreams of a Final Theory*, writes:
> Positivism helped to free Einstein from the notion that there is an absolute sense to a statement that two events are simultaneous; he found that no measurement could provide a criterion for simultaneity that would give the same results for all observers. This concern with what can actually be observed is the essence of positivism. Einstein acknowledged his debt to Mach.

And Weinberg continues;
> Despite its value to Einstein and Heisenberg, positivism has done as much harm as good.

One would be very hard pressed to find the good, and it's precisely in the hands of Einstein and Heisenberg that positivism has done its most significant and lasting damage.

- ♣ You're discarding the very foundation of modern physics.
- → No, just discarding a deficient view of reality.

  As you seem to be very close to conceding, when we rid physics of positivism, we destroy any supposed SRT credibility. And you're right. But discard the foundation of physics? Nah. Sweep away the silt, sand and assorted debris that is SRT and the very solid foundation of all that we know is finally revealed. That's what we're doing.

- ♣ And you do that by relaxing the constraints on what is considered real, on what is allowed as scientifically meaningful. You're opening yourself up to meaningless, ad hoc speculation.
- → Consistency, cohesion, convergence, clarity, concreteness. These are much more meaningful constraints than simply denying reality to anything not subject to direct measurement. And the positivists are the ones so obsessed with what's real and what's not that they end up with no discernable reality at all.

  Direct measurement is necessarily constrained by available tools and is simply not adequate to the task of determining what's real. Reality is not so simplistic. When a proposed reality gives improved consistency, stronger cohesion, better convergence, enhanced clarity and a more rigorous mathematics I don't hesitate to ignore calls for rejection based on inability to directly measure. And if you really want to see meaningless speculation just look at the leading edge of current thinking — examine a little more closely the very obvious results of constraining reality to directly detectable measurement.

  In rejecting positivism we require a much more rigorous science. We're more than compensating for any supposed loss of rigor at a metaphysical level by insisting on much greater discipline where it really makes a difference.

♣ How's that?

→ Inability to measure, may, and probably usually does, mean there's nothing there to measure. But just as surely, it may mean nothing more than inability to measure. Positivism absolutely prohibits the second option. And based on what? Certainly on no known law of nature. Nature's not constrained by any conceivable natural law to reveal all to direct measurement. Can we, then, arbitrarily impose such a constraint? Positivism not only says that we can, but insists that we must, with the curious further assertion that it's the nature of science itself that requires that we do so.

In any case, positivism can't tolerate going beyond measured values to find explanation. We, on the other hand, don't hesitate to go beyond measured values to find good explanation. But, again, what is the nature of good explanation? Our *good explanation* is far more rigorous than positivism's fiat; any good explanation must give us the consistency, coherence, convergence, clarity and concreteness that we always demand.

So, you tell me, which gives the superior science?

♣ And to get this superior science, you're returning to an absolute time and embracing an asymmetry in the theoretical structure that Einstein found intolerable?

→ You got it. Again, simply giving up the arbitrary constraints of positivism allows improved consistency, stronger cohesion, better convergence, enhanced clarity and a more rigorous mathematics. Einstein couldn't tolerate the violation of positivistic principle — so what? Must we allow acceptable explanation to be sacrificed on the altar of clearly dubious principle?

So, yes, we return to an absolute, universal time. Experimentally, we're limited to only relative measurements. We know that — Newton himself clearly recognized that. And we know why we're limited, and why we get the measurements we do. And that understanding in no way conflicts with our intuitive concept of absolute time.

Einstein notwithstanding, there's nothing in modern data that makes Newton's claim with regard to absolute time any less valid today than when he wrote it three-hundred years ago. Indeed, the evidence that he was right is much stronger now than it was then. Newton[44] wrote:

> It may be, that there is no such thing as equable motion, whereby time may be accurately measured. All motions may be accelerated and retarded, but the true, or equable progress, of absolute time is liable to no change. The duration remains the same . . . whether the motions are swift or slow, or none at all . . .

In fact, a change in physical rate of change, regardless of scope (whether local, universal, with regard to only some physical change, or with regard to all physical change) inherently says nothing, absolutely nothing, with regard to the concept of a universal flow of time.

- ♣ As Mach pointed out, you can't even find a scientifically sound definition for your absolute time. So how can it be allowed?

- → Again, in this context your 'scientifically sound' is the essence of positivism. You want direct detection of absolute time? You'll never find it. But if you care about consistency, coherence and convergence with regard to all that we know, then you'll find a universal, absolute time essential.

♣ You're discarding Einstein's space-time concepts completely.

→ Oh, come on. We've said that from the beginning. Maybe you're finally beginning to realize that the change we've been talking about really isn't just metaphysical at all.

Anyway, label the change what you will, we are, in discarding positivism, discarding SRT. With the underlying asymmetry the true foundation of physics stands revealed. Questions long disallowed must now be answered.

We can now sensibly ask why the rate of change of all things that change with time slows with increased velocity — and arrive at sensible answers.

Sure, experiment shows that what Einstein calls time (measured rate of change) is a function of relative velocity — to which the sensible interpretation would reply: Certainly, the true rate of change of all phenomena that change with time is a function of true velocity relative to the ether, and the apparent rate of change, the *measured* rate of change, is a function of velocity relative to the selected reference frame. Nothing strange about any of this.

Let me emphasize again that we have no problem with the measured values or with the mathematics; both are consistent with Lorentz's (and our) ether-based interpretation. And with the MLET interpretation, we can explain, clearly and in a remarkably intuitive manner, why the measured effects are what they are, and why and how the rate of change of all phenomena that change with time change more slowly with increased velocity.

Let's see you do all that using Einstein's interpretation.

♣ Much more than simple passage of time is involved. If things are suddenly so intuitively sensible, what can you now say about simultaneity? Steven Weinberg[45] wrote, regarding Einstein's rejection of absolute simultaneity:

> He found that no measurement could provide a criterion for simultaneity that would give the same results for all observers.

How does your supposed sensible interpretation handle that

→ The strange manner of thinking that makes current physicists believe this finding supports SRT must be a source of genuine puzzlement to any inclined to ask for objective interpretation of empirical data. It's almost as though the physicist, alone within the science community, is totally unacquainted with the concept of 'indeterminate.'

How could the sensible interpretation have any problem with Einstein's finding? His observation was right on the money; he pointed out an important inability to measure. You think you're seeing a problem only because you're making an unwarranted additional assumption. By what logic are we forced to ascribe meaning beyond the obvious: namely, *we can by no means, by no measurement, determine the precise simultaneity of spatially separated events*. It can't be done. And it's not at all difficult to understand why it can't be done.

In the face of an experimentally verified inability to measure precise simultaneity, Einstein did come up with a reasonable way to use the measurements we can make to solve real problems involving relative motion — but, there's no reason to think that this observer-defined simultaneity is more than a very convenient fiction — pragmatically useful at the observational level.

So reality limits us to a pragmatically defined simultaneity (Einstein's pragmatic definition seems clearly to be the best we can do) that does appear different from different perspectives. But we're not measuring true simultaneity at all — we, quite understandably, can't do that. None of this should lead us to conclude that a sensible single underlying true simultaneity consistent with all of the laws of nature has been ruled out.

As a matter of fact, the retention of an underlying absolute simultaneity allows a sensible, intuitive understanding of all measurements and a visualizable physical reality. Conversely, SRT, gives us a physics that is, in Paul Davies' words, "fundamentally alien to the human mind." Again, where's the grounds for preferring the alien alternative over the intuitive one?

♣ You say that your explanation makes little difference in the mathematics or in the way the mathematics is applied. So where's the significant difference?

→ We've gone over this a number of times. In improved ability to predict. In better satisfying our preferences. In improved understanding.

♣ By throwing out SRT? This is crazy!

→ By throwing away every trace of positivism. *Positivism is not a law of nature, and deserves no place in modern science.* We like explanation. Good explanation provides discipline, strength, and understanding. Again, holding to absolute time, including absolute simultaneity, demands the asking and answering of crucial questions — questions positivism (and so Einstein) can't allow.

♣ But, allowing questions doesn't, in itself, provide acceptable answers.

→ No, it doesn't. But at least it does allow the questions. And Mach's refusal to allow the questions is the real science-killer aspect of positivism. The crucial answers will never be found as long as the questions aren't allowed. Good explanation is impossible without good questions.

♣ And the good explanation gives us the theory superior to SRT? You keep making these outrageous blanket assertions. You're all nuts.

→ Good explanation (explanation exhibiting ability to predict, convergence, coherence, and all of the other preferences) is the life-blood of superior theory.

While it's clear that no experiment can give direct access to a measured absolute time, sensible preferences with regard to suitable explanation can give what experiment can't directly provide.

♣ Modern physics clearly doesn't allow your interpretation. Why do you think that is?

→ Modern theory, you mean. No, of course it doesn't. It can't. Einstein, admittedly influenced by Mach, found explanation that goes beyond measured values intolerable. And so the sensible underlying asymmetry was rejected — good-bye sensible explanation.

♣ And you wonder why your effort depresses me? Your arguments may just be sophisticated enough to confuse the lay public.

→ A good first step. Open confusion is much better than unwarranted certainty. If we really can reach the lay public you may finally be called on to answer some difficult questions.

♣ I keep coming back to Davies' argument that all of modern physics rests on the solid base of Einstein's worldview, a base clearly including SRT's no preferred reference frame, SRT's symmetry constraints.

→ Modern physics *rests*? Yeah, like a 'cat on a hot tin roof' twisting and turning, screaming in agony, trying to make sense of experimental data in the garish light of SRT constraints, only to finally concede that 'the reality exposed by modern physics is fundamentally alien to the human mind.' You keep coming back to the SRT base and I'll keep repeating: SRT can't survive. The very fact that the data and the mathematics are very good should tell us that (given the magnitude and scope of current confusion) the problem can't be with them, and so must be with the underlying worldview.

♣ So you propose to return us to an ether (an underlying asymmetry), an absolute time and (I suppose) an absolute space, and by doing that give us a better physics?

→ Exactly. I see the hour's up. Tomorrow, same time, same place?

— *** —

Bemused, David watched Steven leave the room. Then, shaking his head vigorously in an apparent attempt to restore his senses, he entered my office with a dispirited smile on his face. "And you people wonder why you're accused of being modern flat-earthers. You know, the only justification I can imagine for your making such ridiculous claims is that you (I mean all of you) are determined to get Einstein. That bothers me. It's not a good motive and it's especially offensive when you pretend to do it in the name of science."

"You sound like Davies with his '*Why Pick on Einstein?*'

article. 'Get' Einstein? 'Pick on' Einstein? You guys're really something else. You miss the point entirely — SRT is wrong, and it would be just as wrong if it had been Newton, Galileo or Joe Blow who came up with it. A theory's rightness or wrongness has nothing to do with who the author happens to be.

"Einstein's your problem, not mine. Your inability to accept any possibility that Einstein might have been wrong in any significant way seems to me to have nothing to do with science. We can challenge Feynman, Lederman, Dirac, Wheeler, Thorne, Weinberg, Hawking, almost any other highly honored physicist without anyone thinking that we're on a personal vendetta. People may think we're crazy, but they'll at least concede our right to question.

"But challenge Einstein's worldview on any significant point and the immediate assumption seems to be that the motivation couldn't possibly be a simple search for truth, but can only be a driving desire to discredit *him*. I used to wonder what we were doing wrong in presenting our views, but finally concluded that it wasn't us at all. It seems to arise from the mindset of the listener. Einstein is seen as the God of physics, and only a true infidel, an enemy of the one true God, would dare to challenge God's word."

"Come on. 'God of physics.' Where do you get such nonsense!"

"He is the one," I replied, "who, in Lanczos' words, *was never deceived by appearances and his findings had to be acknowledged as irrefutable.* Surely, no self-respecting scientist would ever use such language when referring to a mere man."

"One man exaggerates a little and you make a federal case out of it."

"One man? Get real. The consensus community has no problem with such statements — one finds comparable claims throughout the literature. Look at James Gleick's[46] statement:

> There will never be another Einstein. . . . Einstein's genius seemed nearly divine in its creative power: he imagined a certain universe and this universe was born.

"Even if that's true, why make such a big deal out of it?"

David asked.

"I don't. But as your 'get Einstein' charge proves, you people do. And this cult of the personality is perhaps one of the most damaging aspects of the current situation with regard to any serious search for truth. Why should I care? To the true believer only two kinds of people," I suggested, "disagree with God — idiots or infidels. Until I realized that, I had trouble understanding why disagreement with SRT always evoked either knowing smirks or 'Why are you out to get Einstein?' type questions. Any challenge not obviously stupid is seen as a personal vendetta. We're not," I stated emphatically, "out to 'get' Einstein. Einstein was human — that's the truth."

"OK, enough" David reluctantly conceded. "There probably is some emotional element in my 'get Einstein' charge. But, having conceded his humanity, how about you telling me what you think of Einstein on a personal level. Tell me how *you* feel about the man."

"I'd much rather focus on the science. I don't want my thinking to be judged on the basis of what I think of some supposed superhero. How do I feel about Einstein? I don't see that as relevant to any discussion of reality. What do I think of his ideas? That's an appropriate question."

"Fine. What do you think of his ideas?" David was insistent.

"I'll start with something Banesh Hoffmann[47] wrote:
> Practically all of the basic mathematical formulas of Einstein's 1905 paper on relativity are to be found in the 1904 paper of Lorentz and the two papers of Poincare, both of which latter warrant the date 1905 even though the major one did not appear until early 1906. The presence of often-identical formulas was almost inevitable, since relativity is intimately linked mathematically to Maxwell's equations and the mathematics of wave propagation. Indeed the mathematical transformation that is fundamental in relativity — a formula to which Poincare in 1905 gave the name *Lorentz transformation* — had already been found by the Irish-born physicist Joseph Larmor in 1898

on the basis of Maxwell's equations; and an almost identical transformation had been found by the German physicist Woldemar Voigt in a study of wave motion as early as 1887, the year of the Michelson-Morley experiment.

These things, unfortunately, need to be said because the mathematical similarities have misled some people into the belief that Einstein's contribution was marginal, which it certainly was not. Yet in fairness we must add that among the writings of Poincare one finds so many of the relevant ideas that, with hindsight, one is surprised that he failed to take the crucial step that would have given him the theory of relativity, so close did he come to it.

"Sorry about the length of the passage," I apologized. "I didn't want to bore you, but —"

"No problem."

"But it's important to understand the situation at the time of Einstein's introduction of special relativity."

"You're not," David asked, "going to be one of those who claim that the similar mathematics shows that Einstein's contribution was marginal, are you?"

"Certainly not," I assured him. "When we reject SRT we do reject Einstein's contribution — but that doesn't change the fact that SRT was wholly Einstein's creation. His contribution *was* SRT, the rejection of Lorentz's underlying asymmetry.

"Almost without exception, the other contributors to the mathematics of SRT refused to accept Einstein's interpretation and never accepted SRT as a valid explanation of velocity effects. They continued to insist on an underlying asymmetry. MLET is based on their work, not on Einstein's."

"So we're back to the superiority of MLET?"

"A good place to be." I replied.

"But SRT wasn't Einstein's only contribution. What about general relativity?"

"I don't want to anticipate Mitchell too much in that regard," I cautioned, "so let me just say that the mathematics of general relativity is an impressive achievement, good enough to survive almost unchanged in MLET. The interpretation, however, won't survive."

"In spite of your protests," David insisted, "I think I do detect some personal feeling. If you had to judge Einstein (the man, not the theory) purely on the basis of SRT, what would be your verdict?"

"Again, the theory, I'll judge. The man, I didn't know."

"Just answer my question."

"He was human, he was wrong."

"Just wrong?"

"Ok, from my perspective today (forgetting the turn of the century general infatuation with positivism) his elevating absolute symmetry to universal law, was reckless, ridiculous, incredible."

"And what does that make those like me who believe he was right?"

"You mean those like you who *know* he was right? Very well educated, unable to question settled belief, intellectual followers — lazy, unthinking, naive."

"Be serious."

"You don't think I'm serious? Grow up, lighten up, live a little, trust your common sense, then dare to re-examine those assumptions that so clearly present Nature as absurd."

He paused, smiled, shrugged and turned away — and abruptly turned back. "Look, let's be realistic here. All of you have at least a BS in physics. Big deal. You switched to psychology for your master's, Mitchell spent his whole working life in the computer sciences, and Price got his Ph.D. in electrical engineering. I, on the other hand, got my BS from Stanford at 20, a Ph.D. at 23, worked at SLAC for a year, spent two years in postdoc research at Cambridge, and spent the last few years at CERN. Why do you suppose I don't think you people can educate me with regard to fundamental concepts?"

"I see your point. Maybe we can question Einstein, but doubt David Rhodes? — even I can see that that's going too far."

He left the room without as much as a backward glance.

The Ether
[Steven Mitchell and David Rhodes]
15 May, 1999

→ Yesterday we talked about the differences between SRT's and MLET's concept of time. Asimov asserted that the slowing (with increased velocity) of all things that change with time is, as all positivists must claim, the same as saying that time itself is slowed. We rejected that claim, and argued that the implications of the better interpretation demand a very different worldview, a worldview free of all positivistic influences.

We've made claims. In order to support those claims, we must introduce an ether.

♣ So here comes the ether.

→ As we pointed out the first day, Einstein[48] claimed that,
> Since the special theory of relativity revealed the physical equivalence of all inertial systems [absolute symmetry at all levels] it proved the untenability of the hypothesis of an ether at rest. It was therefore necessary to renounce the idea that the electromagnetic field is to be regarded as a state of a material carrier. The field thus becomes an irreducible element of physical description . . .

God's word to you, perhaps, but absolute nonsense to me — just SRT's gross distortion of any possible understanding of natural phenomena. So you'll need to continue to suspend your belief that a concept should be rejected just because of a conflict with SRT.

♣ If I remember my history correctly, considerable effort

was expended over several decades back in the mid 19th century to define an ether that was stable and could account for the known characteristics of the propagation of light.

→ You're right. And, as one would expect, ether stability and accounting for the known characteristics of all observed phenomena are essential to MLET's ether. But let's leave that discussion for later.

♣ And what about the Michelson-Morley proofs?

→ It's funny, really. Any proposal to return physics to the naive, archaic ether concepts discarded so long ago must seem ludicrous, even to the least enlightened layman; yet the modern theorist can freely ascribe to the 'vacuum' any desired attribute with little offense to anyone. The modern vacuum is acknowledged to be strikingly non-vacuous and has been assigned some rather bizarre attributes — but an ether?

We're stuck with the ether — so let's deal with it.

David Lindley[49], in his book, *The End of Physics*, is more charitable than most when, in discussing the implications of the Michelson-Morley results, he writes:

> But the ether was by now a strikingly odd contrivance. . . . the interaction of matter with the ether had to have the peculiar property of altering the lengths of measuring sticks in such a way that any experiment designed to detect motion through the ether would be defeated and give a result indistinguishable from what would occur if the observer were truly at rest in the ether. The last hurrah of the mechanical ether models was thus a wholly unsatisfactory arrangement: the urge to preserve the stationary mechanical ether was so strong that an unexplained interaction between ether and matter was dreamed up with the sole purpose of

thwarting any experimental attempt to detect it.

And he continues:
> . . . mechanical models of the ether were bound to fail, unless one was willing to introduce some arbitrary extra ingredient such as the Fitzgerald contraction.

♣ Sums up the problem pretty well, doesn't he?

→ Doesn't he though. Terribly embarrassing. Lindley is absolutely right — without the Fitzgerald contraction such models were indeed bound to fail.

And can you imagine? Dreaming up a concept "with the sole purpose of thwarting any experimental attempt to detect it"? How humiliating to have been taken in by such machinations. Without Lindley's insight I'd actually supposed it was introduced because it provided consistency, coherence, convergence, clarity and concreteness. And look at who did this — Lorentz, Poincare, Larmor, Fitzgerald and others, all rather bright men — who would have dreamed? We really should get the word out; because in spite of such a disgusting ploy some of these guys are still seen to have been credible physicists.

Really, how were we to know? Lorentz's Nobel prize has, to this day, never been withdrawn. What's going on? — how could honorable men ever forgive any scientist's dreaming up a concept "with the sole purpose of thwarting any experimental attempt to detect it." Not only that, this particular concept involves a blatant violation of positivistic principles.

♣ Come on — surely you can make your point with a little less dramatization.

→ Make my point? What point can I make? Lindley's contempt, while devastating, is kindly compared to others. Yet . . .yeah, I guess I would like to try.

Dirac[50] wrote:

> Lorentz succeeded in getting correctly all the basic equations needed to establish the relativity of space and time.

I'm puzzled. Dirac, a recognized giant in the development of modern theory, assured us that Lorentz developed "all the basic equations needed" within the confines of an ether theory, a theory that required the Fitzgerald contraction. How strange — a really marvelous coincidence? Should we believe that that kind of coincidence has no significance?

♣ Get on with your . . . whatever.

→ Harold Fritzsch[51] (who held a chair in theoretical physics at the University of Munich and a research position at the Max Planck Institute for Physics and Astrophysics) in his book, *An Equation that Changed the World*, writes:

> We know that protons are extended objects, which we can view as little spheres . . . Now let's make a proton moving at almost the speed of light collide with another proton, or with a nucleus. What happens in this collision is quite complex, and I won't go into it here. But we do know that specific details of this process will be different if the impinging proton looks like a sphere or a disk.
>
> Some experiments of this kind were carried out at CERN. The results were unequivocal. Protons do behave like disks, and the faster they move, the flatter they get, just as [Einstein] predicted.

Just as *Einstein* predicted? It sure sounds like the Fitzgerald contraction verified. Do you think Lindley might be able to help me here? He rejected the ether because

> . . . the ether was by now a strikingly odd contrivance. . . . the interaction of matter with the

ether had to have the peculiar property of altering
the lengths of measuring sticks.

And now, when experiment verifies that the contraction is real, it's *Einstein's* prediction. It's an intolerable ad hoc artifact for Lorentz and a brilliant leap of intuitive understanding for Einstein. Why's that?

Yes, an ether requires a contraction in the dimension of velocity as Lindley pointed out, but so does SRT. What's going on?

Eddington[52], that great early advocate of SRT, wrote in 1918: "When a rod is started from rest into uniform motion, nothing whatever happens to the rod." Now Fritzsch writes;

The results were unequivocal. Protons do behave
like disks, and the faster they move, the flatter they
get, just as [Einstein] predicted.

And Lorentz[53], with his ether-based interpretation, wrote in 1921: ". . . there can be no question about the reality of this change of length."

Fritzsch seems to favor the ether-based explanation, but clearly ascribes all credit to Einstein. What gives?

♣ Give me the rest of the hour and I think I can explain why both Eddington and Fritzsch are right, and why Einstein does deserve the credit for our current understanding.

→ Oh, I won't waste your time like that. We know that 'real,' to the SRT advocate, means little more than 'consistent with the results of direct measurement.' One can't even ask whether the contraction is real in any physical sense. In any case, it seems that Lindley's contemptuous dismissal of real contraction of matter is unwarranted since SRT also requires 'real' (whatever that means) contraction. Either interpretation requires the experimental results.

Actually, the ether-based view tells us that there is an inherent ambiguity in the observations — from within the ether there is simply no way to get an undistorted view of the underlying reality. We can understand that there is (relative to the ether) a real velocity and a real contraction in the direction of that velocity. But we can only measure an apparent velocity (relative to a selected reference frame) and a corresponding apparent contraction. Complex, yes, but intuitively understandable. And it sure beats simply throwing up your hands and declaring that the measurements are the reality — and then simply accepting the derivative mess.

♣ SRT, when applied consistently, gives us the right answers.

→ It does indeed, but, as we cautioned earlier, don't ask what 'consistently' really means. And don't forget where the mathematics came from.

♣ The mathematics are the mathematics of SRT.

→ The mathematics are the mathematics of the Lorentz ether theory. Einstein's contribution was nothing more than applied positivism — no natural imperative involved.

♣ Einstein's explanation is the base of modern physics, a tremendously successful science.

→ Lorentz explained — Einstein simply "listened with supreme devotion to the silent voices of the universe and wrote down their message with *unfailing certainty*." If agreement with things that can be measured is all that you ask you might be inclined to agree with Einstein — provided you're unaware of the better alternative. We do ask for more. We demand improved agreement with things that can be measured. We further demand a consistency, coherence and convergence hopelessly lacking in SRT — and so find SRT unacceptable.

♣ Again, what about general relativity?

→ Throw away SRT constraints, substitute an elastic solid ether for space-time, retain the essential general relativity mathematics, and voila! — a beautifully convergent, new understanding of supposed general relativity concepts.

♣ That, I'd like to see.

→ Stay with us and you will.

But enough of this. We still have a consensus to deal with.

Everyone knows that the Michelson-Morley experiment proved that there was no ether. Every physics book that deigns to admit that there ever was an ether concept says so. Every modern physicist who writes for a popular audience makes the same claim: Michelson-Morley proved there was no ether. Period.

But . . . somebody forgot to tell Mother Nature.

It's interesting to see how one's beliefs shape one's interpretation of concepts and events. I've repeatedly noticed that one finds, often in the same book, both expressions of extreme admiration for the brilliance of Einstein's intuitive leap in claiming the absolute equivalence of all inertial frames and scornful condemnation of the far more conservative notion that matter and all physical processes might be affected by movement through a real physical medium.

♣ Just a second. You persist in talking about 'absolute' equivalence of inertial frames. Einstein talked about 'physical' equivalence, never 'absolute' equivalence.

→ We're talking about physics, and in this context I would argue that Einstein meant to, and succeeded in, making 'absolute' physical equivalence of all inertial frames a fundamental principle of physics theory. I'll continue to use the term 'absolute' to emphasize his obvious personal intent to exclude the possibility of any alternative.

♣ Einstein never used such language.

→ His intent was clear. The term is justified. Modern theory doesn't allow anything less.

As we pointed out a few days ago, the modern physicist's firmness of conviction in holding to the revealed truths of standard relativity dogma is spectacularly protected, at least as far as true believers are concerned, by its acclaimed non-intuitive nature. The correct view of reality, of fundamental truth, is known to be, as Davies and Gribbin[54] write in *The Matter Myth,* "fundamentally alien to the human mind, and defies all power of direct visualization." Ipso facto, any intuitively understandable explanation, any sensible explanation must be rejected. True enough, without culture, one finds paradox — but within the culture it is clearly recognized that any supposed paradox simply arises from asking too much of theory.

From the more conservative perspective this might seem merely a strange aberration, but surely, no one can think the resulting modern mess amusing.

In any case, Michelson-Morley did *not* prove that there was no ether. Quite the contrary. Again, it is, as Dirac and many others have noted, a matter of record that the fundamental core of the mathematics of special relativity was first derived, not by Einstein, but by such men as Lorentz, Poincare, Larmor, and Fitzgerald, to explain the Michelson-Morley results based on an ether model. And, with all due respect to Lindley, whether one subscribes to Einstein's interpretation or the Lorentz-Poincare-Fitzgerald interpretation, it takes some real sleight of hand to avoid conceding the experimental reality of the Fitzgerald contraction.

As Herbert Ives[55] (the Ives of the Ives-Stillwell experiment, the first experiment that directly measured the slowing of clocks with velocity, an experiment which provided significant proof that clocks do slow with velocity in accord with relativity predictions) wrote in 1948:

> The frequent assertion that the Michelson-Morley experiment abolished the ether is a piece of faulty logic. When Maxwell predicted a positive result from the experiment he did so on the basis of *two* assumptions; the first, that the light waves were transmitted through a medium, the second, which was not realized until pointed out by Fitzgerald, that the measuring instruments would not be affected by motion. The null result of the experiment proved *some* assumption made in predicting a positive result to be wrong. The experimental demonstration of the variation of measuring instruments with motion, in exactly the way to produce a null result, shows that it was the second assumption alone that was wrong; leaving the evidence for a transmitting medium, as derived from aberrational and rotational phenomena, as strong, if not stronger, than ever.

It must seem strange to a modern physicist that someone who so completely rejected Einstein's absolute equivalence of inertial frames should be the first to directly prove by experiment that clock rates do slow with velocity in substantial agreement with Einstein. Virtually all modern physicists hold that such slowing proves Einstein's view is correct. Ives, on the other hand, while clearly scornful of Einstein's absolute equivalence, would have been very surprised at any other result — the sensible interpretation requires it.

And Einstein[56] himself wrote:

> Concerning the experiment of Michelson and Morley, H. A. Lorentz showed that the result at least does not contradict the theory of an ether at rest.

Again, what Einstein[57] did believe is clear. He wrote that:

> Since the special theory of relativity revealed the physical equivalence of all inertial systems, it proved the untenability of the hypothesis of an ether at rest.

The ether *is* incompatible with the absolute "physical equivalence of all inertial systems," the absolute symmetry, of special relativity. It is Einstein's absolute, his absolute equivalence of all inertial frames, not the Michelson-Morley results, that required the rejection of the ether — and Einstein's absolute is a poor substitute for a consistent, cohesive and convergent understanding of our physical reality.

As Ives observed (long after he himself proved that clocks do slow with increased velocity) experimental data leaves the argument for an ether "as strong, if not stronger, than ever." The real changes in clock rates and the real contraction of matter seem to require only that elementary particles be composed of interacting distortions of the ether itself, interacting fields within the ether — MLET's three-dimensional standing-wave structures. This requirement is in substantial accord with modern understanding. Steven Weinberg[58] in describing the modern understanding of particles of matter writes:

> All these particles are bundles of the energy . . . of

various sorts of fields. A field like an electric or magnetic field is a sort of stress in space, something like the various sorts of stress that are possible within a solid body, but a field is a stress in space itself.

One need only substitute 'ether' for Weinberg's 'space' to appreciate that the Michelson-Morley results, far from disproving an ether, gave us the first experimental requirement for the current understanding of the nature of matter.

♣ Quite a lecture. I'd like to see you justify that last claim.

→ For now, just let me restate it: From the MLET perspective Michelson-Morley was the first experiment to require that elementary particles of matter be nothing more than interacting ether distortions. The observed contraction of matter and slowing of clocks is an inherent consequence of the nature of matter — stable structures composed of encapsulated, interacting, resonant energy fields — MLET's standing waves. Such particles always exhibit characteristics consistent with measured values (the Fitzgerald contraction and the slowing of clocks) due to the necessity of maintaining a binding resonance, a resonance required for particle stability — a necessary geometrical adjustment to velocity. Again, the measured values are completely consistent with the theoretical necessity.

With increased velocity relative to the ether, this geometrical adjustment within all matter assures that all clocks will slow and all rulers contract in a self-consistent manner. And these effects assure that (as Ives and others have shown) any measurement of the speed of light must always yield two-way values consistent with the true speed of light relative to the ether itself — no matter how the speed of light is measured. Inertial-frame-dependent factors exactly cancel due to self-consistent clock-rate and ruler-length changes. There's nothing strange going on — the overall results are intuitively understandable as velocity-induced standing-wave resonance-pattern changes within particles of matter, within the ether itself.

♣ That's a bit much. You'll need to provide some pretty definitive support to be very convincing.

→ One nice thing about MLET is that, since it requires an intuitively visualizable reality, everything I've just described is easily modeled with computer simulations. Before you leave today I'll give you some papers supporting the theoretical claims, and in the next week or so you'll see the supporting simulations.

Let me just say that, contrary to popular belief, the fact that local measurements always seem to indicate that light speed is isotropic with respect to the observer, far from disproving an ether, demands it. An ether theory, and only an ether theory, can provide a clear, intuitively sensible, consistent, cohesive, convergent and mathematically rigorous, physical explanation of the measured values. Einstein's relativity does not and can not.

Common sense — in every sense — prefers an ether.

♣ The Lorentz interpretation was concurrent with SRT so if what you claim is true why was the ether rejected in favor of SRT?

→ There were a lot of factors. The experimental verification of some predictions of general relativity in 1919 probably was a crucial turning point. General relativity was good enough to convince many previous skeptics that Einstein really did know what he was talking about.

♣ But you don't think that verification significant today?

→ The mathematics of Einstein's general theory is almost identical to the mathematics of MLET's elastic solid. And, as Thorne pointed out, MLET's 'flat space-time' model makes the same predictions with regard to direct measurement that general relativity makes. Would general relativity have had the crucial impact it did, if there had been an awareness of the flat space-time option? Who knows? What we do know is that, with the flat-space-time model, the MLET advantage is a sensible reality that allows predictions that neither the general relativity interpretation, nor the mathematics alone, could ever give us.

♣ So general relativity saved special relativity?

→ Although some think it may have been the decisive factor, it probably wouldn't have been enough by itself to effect such a radical change — other factors played a significant role. Common sense was out of favor, positivism was in favor. In fact, positivism was at its zenith, in terms of widespread acceptance, around the turn of the century — right when Einstein required just such a choke-hold on sensible inquiry. Secondly, the possibility of an unrealized fourth dimension was also a subject of widespread speculation among mathematicians, and also in popular literature.

And it's difficult to completely dismiss social influences from outside the physics community.

♣ That's a strange comment. What do you mean?

→ I don't want to belabor the point but there seems to have been a common intrigue, a fascination, with ideas supposed to exceed ability to understand, among not just the general public, but especially among the elite sophisticates of the early decades of the century. Let me give just one example and then let's move on. Lederman[59], in introducing *The God Particle*, gives a full, otherwise blank, page to a quote he attributes to D. H. Lawrence — it reads as follows:

> I like relativity and quantum theories because I don't understand them and they make me feel as if space shifted about like a swan that can't settle, refusing to sit still and be measured; and as if the atom were an impulsive thing always changing its mind.

Though the true impact is impossible to measure, this attitude with respect to understanding surely influenced many. And you find this inclination to embrace the occult even today, and even within the physics —

♣ Occult? Embrace the occult? Where'd you get that?

→ Occult: Not revealed. Secret. Not easily apprehended or understood. Abstruse. Mysterious. Hidden from view. Concealed. These are attributes that have a definite appeal to many minds.

Would Lederman have used this quote as the introduction to his book if he didn't think it would appeal to many of his readers? And we find the same inclination to emphasize this aspect of current theory in much of Davies' work — just read the chapter titled "Confessions of a Relativist" in *The Matter Myth*. And even Feynman, after urging his listeners to "accept Nature as She is — absurd," goes on to say "I'm going to have fun telling you about this absurdity, because I find it delightful."

Lorentz would deny him this delight. In any case, the ascendancy of positivism, the popularity of a proposed fourth dimension, general relativity and the mystique, the perception that Einstein exceeded rational limits (which he did) seem to be the things that won the day.

♣ And, of course, there was the Michelson-Morley results.

→ Actually the notion that the Michelson-Morley results were in any way decisive in the ultimate rejection of the ether is something of a myth. As noted earlier, by 1904 Fitzgerald, Lorentz, Poincare and others had succeeded in giving a remarkably satisfactory explanation of the unexpected results. But it was an explanation based on an ether.

I think that had a modern understanding of the nature of matter been around at that time the ether-based concepts would have prevailed. And if the Sagnac effect had been demonstrated prior to SRT, SRT probably wouldn't have had much of a chance.

♣ What about $e=mc^2$?

→ We're discussing why SRT won out over Lorentz's ether based theory. Not to diminish the import of the equation, there's little evidence that it influenced the turn to SRT. Prior to Einstein it had been recognized that radiation exerted a measurable force on impact with matter and, while not in the exact form, it had been shown that an equation equivalent to $e=mc^2$ was true for electromagnetic radiation. To the extent that it was understood that matter could be converted to radiation, and radiation to matter, Einstein's conclusion was not terribly surprising. And not to diminish Einstein's contribution in this matter, there is little evidence that he considered it among his greatest contributions. Furthermore, some of Einstein's contemporaries ascribed the first rigorous derivation of the matter-energy relation to Planck in 1907. Ives[60] was one of these.

> Ives argued that:
>
> The equality of the mass equivalent of radiation to the mass lost by a radiating body is derivable from Poincare's momentum of radiation (1900) and his principle of relativity (1904). The reasoning in Einstein's 1905 derivation, questioned by Planck, is defective. He did not derive the mass-energy relation.
>
> And, in the same article, referring to the 1907 study by Planck, Ives wrote:
>
> This derivation [Planck's] of the relation . . . is historically the first authentic derivation of the relation.

In any case, it couldn't have played a convincing role in deciding between SRT and Lorentz's theory, since neither had a problem with the true energy-matter relationships. Of course since Einstein came up with SRT and seems to have been the first to (at the least) explicitly suggest the relationship in this general form, it may have made some difference. But the record indicates probably not much, at least in the early years.

The situation today is different. With almost no modern awareness of the Lorentz theory, most people, both in the lay community and in the physics community itself, associate the equation with Einstein alone. That's unfortunate — as I've said, the Lorentz ether theory had no problem with it.

♣ So it was just fate that made SRT the preferred interpretation?

→ Force of personality, advocacy by outspoken positivists and the sudden interest of leading-edge mathematicians. The positivists and the mathematicians simply preempted the more conservative physicists.

Lorentz, I think, was the superior physicist, but rather than be forceful in the advocacy of his beliefs, he seemed content in his trust that 'truth will out' in the end. He was right, but I'm sure he never dreamed that the Einstein diversion might last as long as it has.

Lorentz[61] suggested in his 1902 Nobel Prize acceptance speech that:

> . . . this leads us to the idea that an atom is in the last resort some sort of local modification of the omnipresent ether . . .

and

> . . . we can therefore never set an electron in motion without simultaneously imparting energy to the ether. . . . in other words, if we determine the mass in the usual way from the phenomena, we get the true mass increased by an amount which we can call the apparent, or electromagnetic, mass. The two together form the effective mass which determines the phenomena.

So we find Lorentz (In 1902) not only suggesting that matter might be very much like what Weinberg[62] describes many decades later:

> All these particles are bundles of the energy . . . of various sorts of fields. A field like an electric or magnetic field is a sort of stress in space, something like the various sorts of stress that are possible within a solid body, but a field is a stress in space itself.

In addition, Lorentz (prior to Einstein) also recognized an increase in effective mass with increased velocity.

♣ That's interesting but I'm not sure I see how it relates to our discussion.

→ Einstein wrote of rejecting the ether because an ether

theory required two types of matter; ponderable matter, and the ether itself. This was a common criticism. With Lorentz's view of an atom as a 'modification of the omnipresent ether,' the ether itself becomes the single fundamental reality, with particles nothing more than stable constructs (Weinberg's bundles) of interacting energy fields within that ether. Einstein's objection is then clearly without merit.

Many mathematicians loved the simple beauty of SRT's (Lorentz) equations. Dirac[63] was extreme in this view of the importance of such beauty. He, at least half seriously, once stated:

> It is more important to have beauty in one's equations than to have them fit experiment.

In any case, the fact that the mathematicians understandably liked the prospect of the mathematics as the essence of physics; and, with the positivists restricting theory to measured values, considerable combined influence was brought to bear in Einstein's favor. (We should note that the true beauty of the Lorentz equations is only enhanced by the way that MLET's asymmetrical reality gives sensible physical meaning to the equations.)

Dirac liked the unquestioned beauty of the perfect symmetry of the Lorentz transformations. Einstein declared this symmetry real and absolute, while Lorentz himself treated it as only an apparent effect of an underlying sensible asymmetry. Again, we like the measured symmetry, and do contend that the mechanics of going from the underlying asymmetry to the as-measured symmetry only adds to the appreciation of the beauty of the equations.

Elie Zahar[64] correctly reminds us [in a paper that is an expanded version of a talk given before the British

Society for the Philosophy of Science on 7 December 1970] that:

> Einstein differs from Lorentz in that he regards the 'effective' variables . . . as the real ones and totally abolishes the Galilean transformation. The [Lorentz's] Theory of Corresponding states is 'observationally equivalent' to Special Relativity [SRT] because experimental results involve only measured, that is, 'effective' quantities. Since the latter satisfy Maxwell's equations, we are unable, whether we adopt Lorentz's or Einstein's theory, to decide on empirical grounds whether our frame of reference is in motion or at rest in the 'ether'.

And:

> As I have already shown, Lorentz's theory is observationally equivalent to the SRT; Einstein's transformed coordinates can be interpreted as the measured coordinates in Lorentz's moving frame. In the latter the 'real' coordinates are still the Galilean ones.

Zahar pointed out in that same speech that an important heuristic rule required by Einstein's interpretation is that one must:

> replace any theory which does not explain symmetrical observational situations as the manifestations of deeper symmetries — whether nor not descriptions of all known facts can be deduced from the theory.

Pragmatically, use of symmetrical mathematical constructs in dealing with measured values is generally warranted by the sensible understanding of real velocity effects, and never requires that the underlying asymmetry be rejected. Einstein's preference was just that: a personal preference based on his youthful positivistic leanings — a preference that demanded that symmetrical observations must always be considered manifestations of real symmetries, a preference that has given modern physics the deserved reputation for being, as Paul Davies tells us, "fundamentally alien to the human mind."

♣ Davies was referring to more than just special relativity. Nature herself is strange. You can't rid modern theory of its strangeness just by replacing SRT.

→ It seems we can. When SRT is removed as a worldview constraint, we can construct a new worldview that's consistent with both experimental fact and the necessary mathematics, a worldview that is also both sensible and intuitive. That, we can demonstrate.

♣ Do you think Einstein might have been able to make an enduring contribution if only he had possessed your insight?

→ Bravo! You're beginning to appreciate the strength of our perspective. Seriously, though, this isn't about Einstein, it's about physics.

Take these papers, review them, and we'll get back together tomorrow morning. The top one is the expanded text of the Zahar speech that I quoted from, and the others are Mansouri and Sexl[65] papers in which the authors, quite rigorously, constructed
> an ether theory . . . that maintains absolute simultaneity and is kinematically equivalent to special relativity.

Tomorrow's Sunday. Let's take the day off and I'll see you back here on Monday. We're ready to look at MLET specifics.

♣ It's not nearly the tenth day. You think I'm ready for this?

→ Probably not, but we'll give it a go. Enjoy the rest of the weekend.

— *** —

I turned off the video recorder and refilled my coffee cup as David entered from the interrogation room.

"Any plans for the weekend?" I asked as he entered my office.

He refilled his own cup before responding. "Not really. Actual theory on Monday? Why now, and not at the start? You had a lot to do with the content and pacing so tell me what you think this week has been about."

"Summarize our aims for this first week? Sure, I'd love to," I said. "Let's take it one day at a time. On Monday our intent was, first, to make one unequivocal, testable prediction, along with strong, but supportable, claims. On Tuesday we wanted to draw attention to the inability of the consensus community to question SRT, an inability not based on solid science but on the elevation of Einstein's claims to 'God of Physics' pronouncements — unquestionable facts.

"Wednesday we aimed to provide a general basis for the objective evaluation of the possible acceptability of scientific concepts. This base was composed of our seven Cs of common sense. The seven Cs are preferences for: consensus, consistency, coherence, convergence, clarity, concreteness, and charm. We then argue that constrictive concepts such as SRT should only be allowed to survive if they *contribute* to compatibility with these preferences. SRT doesn't.

"On Thursday we concentrated on pointing out that the inadequacies of current theory with respect to fundamental preferences has given rise to the 'all we can ask' dumbing down of what is to be considered acceptable. We claimed that we can and should insist on a lot more.

"Yesterday we specifically addressed MLET versus SRT interpretations of the nature of time itself. And this morning we argued in favor of a physically real ether within an absolute space. Throughout, we've argued that all differences between MLET and current theory arise from one unwarranted assumption — Einstein's rejection of a sensible underlying asymmetry. We've characterized the perfect symmetry assumption as 'Einstein's Camel.' So see how much you've learned," I finished.

"Why not the theory first? " David asked.

"You would have," I said, "dismissed it out of hand. You've admitted that you'd never have freely chosen to spend time listening to any concept that clearly conflicted with SRT. That prejudice (with no supporting evidence once positivistic constraints are discarded) is so universal, so strong, that we felt it must be dealt with up front. You needed to appreciate that our disagreement with Einstein is not capricious, is not personal, is in no way destructive of clear understanding, and is based on a return to a more conservative and a more rigorous physics."

The Modified Lorentz Ether Theory
[Steven Mitchell and David Rhodes]
17 May, 1999

→ How was your weekend?

♣ I read the Zahar paper and the Mansouri and Sexl papers you gave me. The arguments seem solid enough as far as they go. It does seem clear that one can explain measured symmetry by introducing an ether and an underlying asymmetry. The question remains: Why would you want to? Einstein's objection still seems valid.

→ Why would we want to? Very good. The whole point of all of our discussions to this point could probably be summed up as our attempt to answer that question. Why choose an ether theory over SRT? As Mansouri and Sexl[65] wrote:

> All experiments can be explained either on the basis of special relativity or by an ether theory . . . This demonstrates . . . the impossibility of an "experimentum crucis" deciding between ether theories and the special theory of relativity.

So direct experiment can't tell the difference. Nevertheless, there is a real difference. As Mansouri and Sexl go on to point out:

> . . . the symmetry group contained in Einstein's theory restricts the possible forms of electrodynamic and other interaction so strongly that on the basis of Lorentz invariance alone an interaction that is known in one system of reference can be rewritten in any other. . . . In other words, the symmetry group contained in relativity makes

many predictions possible, which have to be derived with the help of additional assumptions in ether theories.

♣ Exactly. And SRT gives the right answers in a more straight-forward manner. Occam's razor seems to favor SRT.

→ Even today, this supposed completeness without the need of additional assumptions, is seen as a positive. And, if one doesn't mind a "special" domain (encompassing inertial frames only), if one never asks more than simple ability to predict, one might yet think it so. But we don't like disparate domains, we don't find an absurd Nature preferable to a sensible one, so let's look closer.

First, remember that this 'symmetry group' is in fact the 'relativistic invariance' that, according to Dirac, makes it impossible to be rid of the troublesome infinities of quantum electrodynamics. The ether theory, with an underlying asymmetry, has no such problem.

Second, the necessary 'additional assumptions' do, first of all, require the underlying asymmetry that Einstein found intolerable on positivistic grounds. The underlying asymmetry is reflected in the measured symmetry in such a way that, with the ether theory, one can treat any inertial frame as though it were the true ether frame, but, as Mansouri and Sexl[65] point out,

> By singling out arbitrarily one system . . . to be the ether system one destroys the equivalence of all inertial systems . . .

And that, Einstein could not tolerate. This intolerable asymmetry is the asymmetry that gives us a sensible, physical reality, that gives us the consistency, coherence, convergence, clarity and concreteness that we require — the asymmetry that provides good explanation. SRT, is then, a gross oversimplification that, while superficially appealing, ultimately, unavoidably, leads to absurdity — SRT starts with absurdity and so ends, as it must, in acknowledged absurdity. So, if universal convergence is a reasonable goal, then it's absurd to claim that SRT is the preferred alternative — on any grounds.

Again, from our perspective (as we've tried to say many times before) the best answer to the question of why we should prefer MLET, would be that we like good explanation. And the best explanation is, to us, the one that gives us the greatest consistency, the strongest coherence, the most far-reaching convergence, the most concrete (the most specific and rigorous) mathematics, the most intuitive, visualizable and specific model of physical reality. And with all this, we ask for clarity and charm.

But how do we show the SRT true believer that any of these are sufficiently achievable to give our preferences any validity? If Nature really is absurd, if we really can't ask for more than ability to predict things that can be directly measured, then why should we care about naive preferences?

♣ I suppose that's the essence of my question.

→ I'm not sure I can do better than just show you MLET. But let me try just once more in general terms. Let me describe how MLET became more and more attractive *to me*, the better I understood it.

We've all read the popular accounts of the implications of SRT. For instance, how that when two space-ships traveling in opposite directions pass each other (at a relative velocity that is a significant percentage of the speed of light), each sees the other as shortened in the direction of velocity and each sees the other's clock as running slower than his own. In most such accounts we are told that this situation is non-intuitive and very strange (if not down-right absurd) from the common-sense viewpoint. And, if one holds to the SRT 'no underlying asymmetry' perspective, that's true — common sense must indeed see it as absurd, because there's simply no way that a sensible physical reality can be reconciled with the measured values, *if one holds to SRT.*

Within SRT bounds, that's just the way it is.

But, with MLET's sensible reality — including an underlying asymmetry — the situation is quite different. Now the measured results in the passing-space-ships example — absurd results if one holds to SRT — are in perfect accord with intuition and common-sense. The physical reality so clearly requires the experimental results that even if we had no previous awareness of the observed symmetry, we could derive the mathematics and be forced to conclude that all observation must yield the experimentally verified apparent symmetry so dear to the physics community. With MLET's visualizable physical model the measured symmetry is not at all strange; common sense tells us that the experimental results are required of a sensible reality.

♣ If you can show that, I'd certainly be interested.

→ We can show it. The model is sufficiently specific to allow very interesting simulations of actual physical phenomena. And not just with respect to the particular example. With MLET, the supposed twin paradox mechanism and results are intuitively obvious, the Sagnac effect is likewise intuitive and expected. And this intuitive understanding goes beyond velocity effects. The absence of space-time curvature in the presence of very high homogeneous energy levels is expected. The supposed wave/particle duality of both light and matter is now seen as a reflection of an intuitively obvious physical necessity.

So, with current theory, very many situations (involving not just SRT concepts, but also general relativity and quantum theory concepts) are frequently described in terms such as 'strange,' 'weird,' 'absurd,' 'beyond the limits of human imagination,' and the like. In almost every case, the MLET physical model allows us to respond: "No, not strange at all — a common-sense understanding of what is really happening demands the measured values in a fully intuitive manner."

♣ And all of that by simply retaining Lorentz's underlying asymmetry?

→ That's the single crucial difference between current theory and MLET, yes.

That's really what our discussions to this point have all been about. The point we want you to recognize is that allowing the underlying asymmetry provides a sensible, intuitive, visualizable physical reality from which the current mathematics could be derived and our current empirical base could be predicted. I can't emphasize this point enough.

I think the thing about MLET that first really 'got' me, was that when watching the interactions of the various fields (I had written a few animation programs simulating what MLET told me had to be happening) I could watch the real asymmetrical phenomena produce only symmetrical external measurements as required by SRT. I could watch the intuitively understandable asymmetrical interactions and really appreciate why

SRT, though conceptually dead wrong, would always see the expected measured results within its limited domain of inertial frames. It was even more interesting to see how the underlying phenomena let me move seamlessly into the currently disparate areas of acceleration and gravitational effects, while, again, necessarily deriving only externally observable measurements consistent with current theory — but with a completely different, fully intuitive, conceptual base.

Of course none of this was unexpected given the Mansouri and Sexl treatment of the Lorentz equations and Thorne's 'flat space-time' treatment of general relativity predictions (MLET's ether-based reality, embracing both an absolute space and the universal flow of time). Not surprising, but it was still very interesting to watch how (at very high velocities) the very large asymmetries with respect to so many contributing factors work together to necessarily produce the symmetrical observed values — it's really beautiful. It's especially fascinating to watch how the very large changes in the relationships between, say a proton and an electron, aren't even felt by the particles involved, due to these self-consistent changes. It also becomes very obvious why, in denying the reality of the underlying asymmetry, current theory must see Nature as absurd. This overwhelming beauty will never be realized so long as one holds to SRT's absolute symmetry.

You can't get this type of consistency and convergence with SRT. And Einstein's objection to the underlying asymmetry is made on very dubious grounds, certainly not based on any natural law. All of this is why I have so much confidence that the LIGO experiments will detect nothing — MLET, with its universal convergence, doesn't allow the expected type of wave.

♣ But if that's true, if the ether-based explanation is really as consistent with relativity data, as rigorous as it seems, why hasn't it had more impact? I still have trouble believing that the physics community wouldn't have discovered the better interpretation (if it's real) if they'd ever been so close to it.

→ We discussed some of the reasons Saturday morning. Think about it. Einstein admitted that Lorentz had succeeded in showing that the Michelson-Morley results were not inconsistent with an ether at rest. He merely claimed (in the 1920 speech[66]) that:

> For the theoretician such an asymmetry in the theoretical structure, with no corresponding asymmetry in the system of experience, is intolerable.

It's easy to forget that the Mansouri and Sexl ether theory, as must all ether theories, introduces "an asymmetry in the theoretical structure." Modern physics simply followed Einstein's lead in rejecting this as intolerable. And it clearly, explicitly, was rejected because 'no corresponding asymmetry in the system of experience' was found — not because it made no difference.

♣ But it does make no difference in the predicted values within the scope of SRT.

→ Ah, very good; that's very close to the truth. You're starting to recognize the real, arbitrary constraints that current theory demands. You may be, I suspect, beginning to recognize that, if we allow our concept of reality to require universal convergence, we must recognize that things that make 'no difference to the predicted measured values within the scope' of some arbitrary domain of applicability, may, nevertheless, make a very big difference in our understanding of the totality of our real, universally convergent reality.

In that respect, you can view MLET as the natural outcome of the rigorous eradication of arbitrary domain boundaries. Now, given this understanding, we could, perhaps, agree that we can safely reject concepts that, whether true or false, make no difference. We might, then, agree on what is merely metaphysical. But with one caution: We still don't accept the 'makes no difference' test as a very good one — because who can tell us at what point we can really be sure that an alternative concept makes no difference?

Holding theory to the best possible consistency, coherence, convergence, clarity and mathematical and physical concreteness will always serve much better in our sensibly disciplined search for understanding.

- ♣ You may be starting to make sense.
- → All of this contributes to our confidence that eradication of all traces of positivism is essential to good theory.
- ♣ Hawking notwithstanding, I don't think that positivism is generally held in particularly high regard within the physics community. I hardly think it can still be considered a significant factor.
- → You seem to believe that. I find that strange. Einstein's 'intolerable,' and Born's contempt for concepts introduced to 'explain,' and the generally accepted rejection of the physical reality of concepts not ascertainable by direct measurement, all exhibit the very essence of positivism.
- ♣ But you've already conceded that you could accept the rejection of concepts that make no difference. Does

that make you a positivist?

→ Not at all. I said 'make no difference' not 'make no directly measurable difference.' And I was further careful to insist on the eradication of domain boundaries in considering whether or not a concept makes a difference. Even then I said it was not a good test — there is simply no sure way to ascertain with absolute certainty that alternative explanations 'make no difference.' And, in fact, if alternative explanations are themselves clearly different (i.e., involve mutually exclusive concepts), it is almost certain that in some very real way, their implications must differ.

♣ Even if you're right, so what?

→ The question becomes: "How could Einstein possibly know what he claimed to know?"

Again, there's no law of nature that says that reality, that good theoretical structure, cannot exceed experience, cannot go beyond ability to directly detect. There is likewise no reason to reject good explanation simply because it involves constructs 'introduced just to explain.' After all, 'concepts introduced to explain' is really a pretty good definition of theory itself.

The bottom line is that the consensus community can't allow the alternative no matter how consistent, how cohesive, how convergent and how concrete it may be; no matter how much better it works. Nature simply can't be allowed to hide anything from direct measurement — no matter how sensible the mechanism She uses. Since man is, then, the final arbiter with respect to what's real (in any allowable, scientific sense) Nature must be appropriately constrained by man's limited ability to directly perceive — that reasoning is the real justification of the turn from Lorentz to Einstein. So what are we to think of SRT?

Brilliant, intuitive leap? Naive? Or just plain stupid? You decide.

♣ You're claiming that there are good explanations (explanations just as rigorous as Mansouri and Sexl's ether-based explanation of relativity effects) in very many areas of physics that have been rejected only because of the influence of positivism on modern thought?

→ You've got it. Look . . .

Remember, it's been shown by numerous writers that special relativity effects can be rigorously explained, as Mansouri and Sexl put it by "an ether theory . . . that maintains absolute simultaneity and is kinematically equivalent to special relativity."

And Kip Thorne reminds us of the well known fact that when comparing predictions of general relativity's curved space-time paradigm, and the flat space-time paradigm (fundamentally MLET's space and time) "we can be sure that when the same physical situation is analyzed using both paradigms, the predictions for the results of experiments will be identically the same."

And with regard to quantum theory, Stanislaw Ulam[67] writes:

> I told Fermi how in my last year of high school I was reading popular accounts of the work of Heisenberg, Schrodinger, and De Broglie on the new quantum theory. I learned that the solution of the Schrodinger equation gives levels of hydrogen atoms with a precision of six decimals. I wondered how such an artificially abstracted equation could work to better than one part in a million. A partial differential equation pulled out of thin air, it seemed to me, despite the appearances of derivation by analogies. I was relating this to Fermi, and at once he replied: "It [the Schrodinger equation] has no

business being that good, you know, Stan.

In all three cases, special relativity effects, general relativity effects, and quantum theory's Schrodinger equation, we can find explanations essentially compatible with classical mechanics, with Euclidean geometry — explanations that have, to use Fermi's words, "no business being that good."

Is it just coincidence?

♣ If true, maybe a little bizarre. But does it in any way *contradict* current theory?

→ As we've said many times before, current theory is protected by the 'all we can ask' axiom of the modern consensus community. If you dare ask for more, then look a little closer: If the conservative (MLET) interpretation is true then it is easy to see why the consensus interpretation works as well as it does. The modern consensus view is simply the positivistically constrained reflection of the intuitively understandable conservative perspective. Explicitly, the consensus view is the MLET view forced (to the extent possible) into a positivistic mold.

On the other hand, if Einstein's view, the current consensus view, is the reality, then one is left with a rather profound dilemma: The conservative MLET interpretation simply has no business working at all, let alone as well as it does.

♣ So Einstein's intolerance of an underlying asymmetry in the theoretical structure, his insistence on the absolute equivalence of all inertial frames, is the simple unwarranted assumption, the assumption that only an outsider could question, that turned theory toward inevitable confusion?

→ Exactly. And positivistic constraints have so severely

distorted the science over such a long time that the community has simply lost any inclination to trace the confusion back to initial causes. But get rid of the positivistically inspired constraints and you have a sensible science once more.

♣ Again, if you're right it's almost impossible to believe that none of us have discovered this supposed superior interpretation. No matter what you think, we have all been searching for truth.

→ But where do you look? You firmly believe that truth that conflicts with SRT is not truth. And, as Dingle's experience proves, truth found by asking disallowed questions can destroy a career, but otherwise may have little or no impact.

> And as Einstein[68] pointed out,
>> It is often, perhaps even always, possible to adhere to a general theoretical foundation by securing the adaptation of the theory to the facts by means of artificial additional assumptions.
>
> And Steven Weinberg[69] wrote:
>> There is nothing in any single disagreement between theory and experiment that stands up and waves a flag and says, "I am an important anomaly."

♣ So you're saying?

→ Securing adaptation of new facts to SRT's perfect symmetry, to the "general theoretical foundation" is such a challenging task that physicists have little time or inclination to re-examine the foundation itself — especially given that that foundation is known to be fact. And without a reassessment of the whole, it's fairly easy to discount the very many single disagreements. But the flags are flying.

♣ So tell me about this alternative. Show me MLET.

→ I don't want to exhaust your patience, but let me first review a few crucial points just once more.

As we've said before, I really want to emphasize that the facts (the raw experimental data) and the mathematics supposedly supporting special and general relativity, only gain credibility in the transition to MLET. And modern quantum theory will find profound relief in the lifting of SRT's positivistically inspired symmetry-at-every-level constraints. As Dirac[70] noted with regard to getting rid of the troublesome (to him at least) infinities of quantum electrodynamics:

> One can thus make quantum electrodynamics into a sensible mathematical theory, but only at the expense of spoiling its relativistic invariance.

Although he found a reasonable solution, he just couldn't bring himself to question SRT's invariance demands and so concluded that something must be wrong with quantum theory. Since the problem could never be traced to quantum theory, the current consensus is that, since SRT itself is 'fact,' one must, in order to secure 'adaptation of the theory to the facts,' simply give up rigorous mathematical integrity — so much for mathematics as a sure guide.

Again the *measured* symmetry holds, but only as a reflection of an underlying asymmetry. Current theory gets in trouble by its inability to discriminate between measured symmetry and real asymmetry — it allows no distinction between measurement and reality, and so must lead to absurdity and confusion.

The Modified Lorentz Ether Theory, as the name makes clear, is based on an ether. MLET asserts that all phenomena, all matter, all energy, all of everything — all that we can observe and all that remains hidden — is encompassed in an energy-filled, dynamic, elastic-solid ether.

Strangely, Einstein went so far as to suggest that it was Maxwell who first relieved us of the necessity of an ether by reducing electromagnetic radiation to interaction of fields, fields that, utilizing Einstein's strange magic, somehow became 'mathematically independent' (whatever that may mean) of the necessity of any medium.

Maxwell[71] would surely have been surprised and probably offended at such a claim.

The last few sentences of the *very last paragraph* of the unabridged third edition of his two volume, *A treatise on Electricity and Magnetism*, reads as follows:

> Hence all these theories lead to the *conception of a medium in which the propagation takes place*, and if we admit this medium as an hypothesis, I think it ought to occupy a prominent place in our investigations, and that we ought to endeavor to construct a mental representation of all the details of its action, *and this has been my constant aim in this treatise* (emphasis added).

And we have one thing he was denied — we have a vast storehouse of empirical data and a rich and illuminating abundance of well validated mathematical relationships. I think he would have welcomed the modern MLET ether.

But what can we now say about this ether? You mentioned earlier the problems 19th century theorists had with specific models of the ether. MLET does not pretend to answer all questions with regard to the actual structure of the ether, but it does insist on characteristics consistent with observed phenomena. Stability is required, support of light waves (transverse vibrations, i.e. vibrations of shear strain) is required and absence of longitudinal or compressive waves is required. Since shear waves can only be supported in a solid, the ether might best be pictured as a solid. But all known solids support compressive strain waves, and the ether clearly does not. MLET's author describes his ether as a modified MacCallaugh ether.

We'll look at the actual characteristics in some detail after we move to the lab and watch the simulations of what actually happens. For now let's just note that there are two fundamental types of distortions: rotational (shear) and compressive. Given the appropriate characteristics, neither shear oscillations alone or compressive oscillations alone would propagate, but just remain in place and oscillate. In fact, if they were out of phase, a combined shear and compressive oscillation, would not propagate but simply stay in place and oscillate. Only combined shear and compressive oscillations, oscillating in phase, will propagate.

♣ Doesn't sound like any solid I ever heard of.
→ No, it's clearly different than any known solid. But since all matter, all known solids (as we shall see) are themselves simply constructs of resonant fields (standing-waves) within the ether, it would seem surprising indeed if any should exhibit precisely the same characteristics as the ether itself. And for that matter, how the actual structure of known elastic solids gives them their measured characteristics is not itself well understood.
♣ So with all this you still can't describe in detail the actual structure of the ether?
→ At this point we haven't attempted that, no. For now, we acknowledge that all we really know about the ether are those characteristics that are essential to a sensible accounting for observed phenomena.

- ♣ So characteristics are simply postulated to explain?
- → That's right. But they're very specific characteristics that do yield quite rigorous results.
- ♣ But isn't that rather ad hoc? Where's the vaunted discipline?
- → Remember that, in keeping with our definition of good explanation, the assumed ether characteristics must exhibit in themselves consistency, coherence, convergence and mathematical concreteness — no trivial restriction.

    So you're left with a clear choice — positivistic fiat or good explanation. Take your pick. Just be aware that the good explanation gives the superior ability to predict, the more rigorous interpretation of the mathematics and the better agreement with the experimental data.
- ♣ So show me.
- → I just realized that we've spent a lot more time reviewing previous points than I intended. If we continue we'll run well past an hour just to cover the basics. Would you rather quit now and cover theory tomorrow?
- ♣ I'd rather cover at least the basics now, if you have the time.
- → OK. Let's summarize the new worldview, the MLET perspective.

The Universe

The ether *is* our Universe, nothing less, nothing more. It can be best understood as an elastic solid, though with some characteristics quite unlike any familiar solid. If the universe is expanding it is the energy-laden ether that is expanding.

Energy

All *energy* within the universe is encompassed in (consists of) *elastic distortion of the ether*.

♣ *All* energy?

→ All. And so any elastic distortion can properly be thought of as an energy *field*. All observed and observable phenomena are nothing more nor less than manifestations of interacting fields, the (sometimes very complex) interactions of ether distortions.

♣ So what we have is very many different types of distortions, many different types of fields?

→ In a sense, but that's somewhat misleading. Actually there are only two *fundamental* field types.

♣ Only two? You have some explaining to do.

→ All observed and observable phenomena are composed of no more than two fundamental field types. Compressive fields, density gradient fields, are *gravity fields* and shear or twist fields (created by motion relative to the ether) are *kinetic fields*.

♣ Do you really understand the magnitude of such a claim? "All observed and observable phenomena are nothing more nor less than manifestations of interacting fields"? And there are only two types of such fields?

→ That's right.

♣ So electromagnetic waves, gravitational waves, all forms of matter (both stable and unstable), all forces- the electric force, the magnetic force, the strong nuclear force, the weak nuclear force and gravity itself are all manifestations of these two field types?

→ Interacting in various ways, yes.

♣ And you can handle polarity and all other

- ♣ characteristics within this framework?
- → Quite well.
- ♣ So show me what you can do with these magical fields.
- → The fundamental fields interact, and field interactions may produce resonant (stable) oscillations or transitory non-resonant (unstable) oscillations, and oscillations produce the dynamic fields.
- ♣ And so particles of matter, including all related fields, are completely defined by compressive and shear type fields?
- → That's right. Resonant interactions of the two fundamental field types may produce self-sustaining wave patterns, stable wave patterns, ranging from relatively simple to very complex.

    Elementary particles of matter are three-dimensional, resonant constructs of standing waves — standing waves consisting of *out-of-phase* interactions of gravity and kinetic fields. These standing waves form stable structures in three dimensions.
- ♣ And the associated fields?
- → The structure's wave patterns are such that, to an external observer, the standing waves appear to rotate around the particle. This apparent rotation is the particle's intrinsic spin. And all fields associated with the elementary particles are manifestations of the effects of the dynamics of the particle's standing waves, including this intrinsic spin.

    Consistent with these observations, the interactions of the various standing waves which make up a particle of matter create rotating external phase movements of *compressive* and *twist* oscillations. These phase movements define the particle's *electric* and *magnetic* fields respectively.

- ♣ So you're saying that electric fields are —
- → Electric fields are composed of the rotating phase movements of *compressive* oscillations, *oscillating gravity fields*.
- ♣ And magnetic fields?
- → Magnetic fields arise from *twist* oscillations, *oscillating kinetic fields*.
- ♣ And how do you explain polarity?
- → Both electric and magnetic polarity are determined by the nature (outward-moving or inward-moving) of the phase movements associated with the respective oscillations.
- ♣ And by what magic do you conjure up the gravity field?
- → A particle's standing waves displace internal ether. This internal ether displacement creates an external *ether-density gradient of compression*; this is the particle's *gravity field*.
- ♣ And electromagnetic waves are?
- → The simplest kind of field interactions. Electromagnetic waves are interacting compressive (gravity) and shear (kinetic) fields, *oscillating in phase*. MLET would prefer the term gravitokinetic, since 'electromagnetic' (with the above definitions of electric and magnetic fields) implies the oscillation of an oscillation, which these waves are not.
- ♣ I'm beginning to feel like a straight man for some weird comedian. But I'll play along — what about the various forces?
- → MLET differs significantly from current theory, in that, with an ether, there is no need for force carrier particles. All forces can be intuitively understood as the expected interaction characteristics of the fields themselves, i.e.,

as the interactions of the elastic ether distortions associated with the particles. That is, all forces are directly identified with interactions of the various corresponding energy fields.

♣ So the force of gravity arises from —

→ Intersecting compressive density gradients— *intersecting gravity fields*. Intersecting density gradient fields will create an effective attraction between ether-displacing structures because the total ether displacement is reduced as each body moves deeper into the other's external field. So the gravity force is always an attractive force.

♣ And the electric forces?

→ The electric force arises from the interaction of *electric fields*. An electric field is produced by rotating, oscillating gravity fields, and the nature of the phase movements (outward-moving or inward-moving, depending on the structure of the originating particle) determines the electric field's polarity. Quite sensibly, like phase movements (like charges) repel, unlike phase movements (unlike charges) attract.

♣ And I suppose the magnetic force and polarity are analogous, but related to kinetic fields?

→ Right. The magnetic force arises from the interaction of *magnetic fields*. Again, a magnetic field is produced by *oscillating* kinetic fields, and the nature of the phase movements determines the magnetic field's polarity. Once again, like fields repel and unlike fields attract.

♣ And I'm sure you have a ready answer with regard to the nuclear forces.

→ We're moving closer to pure speculation here, but it seems likely that the nuclear forces arise from physically rotating elementary particles. The physical spin is always in addition to the particle's normal intrinsic spin. These forces cannot normally arise from isolated particles because physical rotation of an elementary particle is probably an unstable phenomenon in the absence of stabilizing forces provided by the very close proximity of other rotating particles — maintenance of the physical spin probably requires the presence of appropriate external forces.

This gives us a strong nuclear force that is a very localized, very intense gravity field — many orders of magnitude stronger than gravity fields produced in the normal manner. It is presumed that this intense field is created when, within a very small volume of the ether, the phase movements of merging electric fields interact and produce a smoothed (non-oscillating) ether density gradient. These are seen as extremely strong, extremely local, compressive density gradients (gravity fields) in some discrete volume of the ether in the vicinity of the spinning particles.

Similarly the weak nuclear force is a very localized shear (kinetic) field. The physically spinning particles produce a very local merging (smoothing) of magnetic phase movements, analogous to the strong force, but produced by shear rather than compression.

And that's MLET's basics.

- ♣ Any reductionist should find it appealing. But, I'd bet, impossibly simplistic.
- → Conservative, sure — but a solid base for intuitively understandable complexity.
- ♣ If it really works, let me ask how you explain some well known facts.
- → Fire away.
- ♣ The speed of light as an ultimate speed limit?

→ With all phenomena based on interactions of ether distortions, the local ether reaction time determines the speed of light. The speed of light is, in a sense, a measure of reaction time and so reflects a definite limit to the speed of all interactions, all waves and all structures composed of waves. Since all matter is composed of standing-wave structures, this limit clearly applies to all matter.

♣ Velocity effects?

→ All matter is composed of resonant standing-wave structures. Resonance is essential to stability. No resonance would be possible if the standing-wave structure were to equal or exceed the speed of light — a clearly impossible feat. And, given any *possible* speed, the required resonance can only be maintained, if for any change in velocity there is a corresponding accommodation of that new velocity by a definite change in the geometrical structure of the moving particle.

♣ And, in this manner, you can explain why we see both a shortening of measuring sticks and a slowing of clocks with increased velocity?

→ Both the slowing of clocks and the shortening of measuring sticks are necessary consequences of the structural accommodation to increased velocity. Furthermore, the velocity effects are such that the measured symmetry, Einstein's absolute symmetry, is always assured by the nature of the underlying asymmetry, regardless of the choice of inertial frames.

♣ The increase in mass?

→ An increased velocity involves a structural adaptation that produces a corresponding increase in effective mass — I say 'effective' because both compressive and shear (gravitational and kinetic) effects change with velocity. Both affect measured values, but in different ways. We'll see this in the simulations.

♣ General relativity's equivalence principle?

→ Let's wait 'till we get to the lab for that. While it's reasonably straight-forward and intuitively understandable, it does require a careful definition of terms, and without the simulations it's hard to convey the necessary distinctions. A brief explanation wouldn't sound that much different than Einstein's assertions, but again, MLET explains, general relativity could never do that, current theory can't do that. MLET's intuitively understandable physical reality is, once again, a very big improvement.

♣ So your definition of inertia itself?

→ *Inertia* is the resistance to change — once a particular velocity is accommodated, that velocity is effectively locked in by the resonance and displacement patterns and, since any pattern change involves a specific corresponding energy change, these patterns can only be changed by an external force.

♣ And gravity? — especially as it relates to general relativity?

→ A gravity field is analogous to Einstein's space curvature. However, since a gravity field is a density *gradient*, no homogeneous density, no homogeneous energy level, can produce a gravity field. Only the presence of some form of encapsulated energy (such as matter) can create a density gradient, a gravity field. This explains why the presence of the extremely high levels of energy everywhere in the 'vacuum' doesn't create the kind of space curvature predicted by general relativity.

- ♣ And how does a gravity field produce the kind of effects required for Thorne's flat space-time model? The model that you claimed is compatible with MLET?
- → The rate of change of all phenomena that change with time is a function of ether reaction time, and reaction time is a function of ether density — reaction time slows with increased density. So any local change in ether density will result in a real change in the physical rate of change of all phenomena that change with time. Since a gravity field is a density gradient (a density that constantly changes with distance from the source) field, all phenomena are affected by gravity. The speed of light slows, and the structural resonance geometry of particles of matter change due to gravity's effect on the standing waves. So rulers are shortened and clocks slow down with the increase in ether density as one moves deeper within a gravity field.
- ♣ You've related particle stability to resonance. Explain that.
- → The key to all stability is *resonance* — without it matter, 'bundles of energy,' could not exist in stable forms because lack of perfect resonance must result in energy loss. So, in the absence of destructive external force, perfect resonance assures enduring stability. Conversely, any lack of resonance introduces some level of instability.
- ♣ And you can get perfect resonance without perfect symmetry?
- → Yes. In fact, given a velocity greater than zero with respect to the ether, resonance *requires* an asymmetry in the dimension of that velocity.
- ♣ You've covered an awful lot. Do you have it spelled out in something I could read at my leisure?

→ Sure. And we're way over the hour. Pick it up from Julia on the way out.

♣ What's in store for tomorrow?

→ We'll move on to some of the elementary particles. Given specific ether characteristics, MLET's most promising early successes were in providing sensible, visualizable models of some elementary particles — models consistent with known particle characteristics.

— *** —

As David stepped into my office I handed him the paper Steve had referred to, *The Basic Concepts of the Modified Lorentz Ether Theory*. He took it, gave the title a cursory glance, and asked, "You really think something this simple, this concise, can provide a comprehensive foundation for all that we know?"

Normally I don't care too much for the really 'far out' speculations of modern physics; and I considered Micho Kaku's book *Hyperspace: A scientific odyssey Through parallel Universe, Time Warps, and the 10th Dimension*, about as far out as a serious physicist could get. But a friend of mine had recommended that I read the book and it was on my desk. David's question reminded me of a quotation Kaku had used.

I opened the book, found the quotation and answered: "Kaku[72], in emphasizing the importance of simplicity and elegance, uses a quotation from Richard Feynman that pretty much reflects my impression of the significance of MLET's basic simplicity. Feynman said:

> You can recognize truth by its beauty and simplicity. When you get it right, it is obvious that it is right — at least if you have any experience — because usually what happens is that more comes out than what goes in . . . The inexperienced, the crackpots, and people like that, make guesses that are simple, but you can immediately see that they are wrong, so that does not count. Others, the inexperienced students, make guesses that are very complicated, and it sort of looks all right, but I

know it's not true because the truth always turns out
to be simpler than you thought.

David smiled. "And why am I not supposed to think you're one of Feynman's inexperienced crackpots?"

"I'd say it's the 'more comes out than what goes in' nature of MLET's basics. In developing these basics the author's initial goal was very modest. At most he'd thought to suggest a new direction of inquiry. What came out was more than he'd ever hoped for in his most optimistic dreams. I like that. As Feynman points out, that's *not* a characteristic of crackpot theory.

"And of course there's the remarkable confirmation of the essential validity of the existing mathematics, coupled with unsurpassed convergence, consistency, cohesiveness and all the rest. No crackpot theory could ever be expected to embrace so convincingly the current mathematical and empirical base that is so important to the caring physicist."

"If MLET were really that consistent with respect to current knowledge," David said, "then it certainly would be appealing. But I doubt very much that that can ever be demonstrated. I'll admit, though, it's much more interesting than I'd ever have guessed."

MLET — Beyond the Basics
[Steven Mitchell and David Rhodes]
18 May, 1999

♣ You've given me a lot to think about.

→ I hope that means you *are* thinking about it.

♣ I'm not ready to doubt SRT, it's just that I find it fascinating that MLET, at a non-specific level (superficially, probably would be more accurate) seems to make sense. Obviously, very many crucial specifics are still missing.

→ It does make sense — good sense. And by that I mean exactly what we discussed last week. Good sense demands consistency, coherence, convergence, clarity and concreteness. And *concrete*, does mean *specific*.

♣ As I said, I'm intrigued.

→ So let's move to some specifics. Let's first look at a very specific model of the electron and suggested explicit structures of a few other particles. It should be understood that while it's expected that the models will be shown to be essentially correct, the absolute correctness of these particular models isn't crucial to the theory.

Elementary Particles. Every stable elementary particle can be pictured as a cube. I have here a very large pair of dice, let's use one as a visual aid. Now, holding the die so that faces 1, 2, and 3 are pointing up (with the common corner pointing straight up) let's call their common corner the north pole, and let's call the three faces the northern hemisphere. Faces 4, 5, and 6 can then be viewed as the southern hemisphere and their common corner as the south pole.

Now let's view the cube as a construct of interacting standing waves; each face of the cube is composed of standing waves which rotate in the plane of the face. These waves are out of phase gravitational (compressive) oscillations and kinetic (twist) oscillations. Due to the three-dimensional interactions, they do not simply oscillate back and forth; rather the waves of each face rotate within the dimensions of the particular face.

This rotating disturbance within each face is two wavelengths long, so that, with the rotation, a compressive maximum will occur simultaneously at opposite corners of the respective face; and maximum expansion (minimum compression) will occur at the other two corners.

The actual direction of rotation and associated phase movements of each of the various faces, when coupled with any motion of the entire structure will determine the nature of the particle's various fields (electric, magnetic, gravitational and kinetic). As we shall see, the spin characteristics of truly elementary particles are completely defined by phase rotations within the structure (intrinsic spin) while other particles (particles making up composite structures such as the proton) may require an additional rotation of the entire structure (physical spin). It seems likely that physical spin can only be maintained within the more complex resonance patterns of composite particles.

Now let's proceed to specific particles.

The Electron. Let's again turn the cube so that the corner defined as the north pole points up (faces one, two, and three facing upward). Now we can begin to picture the standing waves. Remember, each wave rotates in the plane of, and around the edges of, a particular face.

With the north pole up, let's look down at faces one, two, and three, and start with their respective standing waves all having a maximum compression at the north pole corner of the respective faces. With all three faces reaching a maximum compression at the same time, at the same corner, the three-dimensional compression at the pole reaches a maximum. At this time each top-hemisphere face will also have a maximum compression at its bottom corner.

That's half of the picture, the top hemisphere. Now, let's look at the bottom hemisphere. When the north pole reaches *maximum* compression (maximum compression phase of all of the upper standing waves at the pole), all the lower faces will reach *minimum* compression at the south pole. They will also have minimum compression at their uppermost corners. The other two corners will have maximum compression.

Now, if we look at the cube carefully, *with the north pole at maximum compression*, we should note that *every corner* of the cube is at either maximum compression or minimum compression. That is, the three faces touching a corner will all be at either a maximum compression or a minimum compression at that corner. Further, if we look only at the six 'equatorial' corners (the six corners closest to the equator) we will note that the corners alternate, compressed, expanded, compressed, expanded, etc.

Now let the rotations begin. The standing waves of all six faces, when viewed from above (let the cube be transparent so we view both top and bottom faces from above) will rotate clockwise. Here it begins to get interesting. Note that although the structure itself is not rotating, there is an apparent or intrinsic overall rotation at the zigzag equatorial zone as the phases of the adjacent waves seem to move smoothly (albeit up and down also) around the structure, through the six equatorial corners. This is the primary spin of the electron.

Now a cube can also be viewed as three pairs of opposite faces: 1 and 6, 2 and 5, and faces 3 and 4. If we look down through any of the standing wave pairs (faces one and six for instance) we will also see a spin with an axis running through the centers of the pair of faces. These secondary spins contribute secondary effects to the electron's fields.

The overall electron structure does not physically rotate but the intrinsic spins have effects just as real as would any spin of the complete structure.

So the electron is a bundle of oscillating energy with various intrinsic spin components and with the phase of the oscillations moving through the ether at a speed determined by the ether density- the speed of light. The intrinsic rotation assures that energy is not radiated, and the extended oscillating fields simply act back upon themselves.

Again, when the top corner is compressed, the three corners just above the spin equator will be expanded. At the same time, the three southern corners will be compressed, and the south pole will be expanded. The compression followed by expansion of the corners will oscillate in phase while maintaining the opposing relationships described above. But, if each corner is oscillating in phase, there must be a radial change of phase with distance from the center of the structure. It is not difficult to see that this phase may move inward or outward.

An inward or outward movement of phase cannot occur without energy radiation, *unless* there is an apparent spin of the standing-wave structure. Because of entropy considerations, the electron is chosen to have outward moving phase and the positron an inward moving phase. Note that the two poles of the electron will also have outward-moving phase and that the two poles of the positron will have inward-moving phase.

To this point, the description I've given is essentially Ron Hatch's (the author of MLET) description. To give you a more complete understanding of why we like this model, I'd like to now read a paragraph directly from his description[73]:

> It is also apparent from the model that the correct, but unusual, relationship between spin and magnetic dipole moment will result. The three pairs of spinning faces combine as a vector to give the total spin value, but half of the spin is canceled at the common boundaries of the faces. The angular relationships are such that the spin of the total particle is slightly less than the spin of one pair of opposite faces. The spin of a pair of faces is the inverse of the square root of three, while the spin of the entire structure is one-half (in units of Planck's constant over 2 pi). The magnetic dipole of each pair of opposite faces adds as a vector without any canceling effect. Thus the approximate g-factor (very close to two) of the electron is obtained directly. The magnetic dipole moment is twice as big as it would be if the structure were spinning as one combined whole.

As you can appreciate, all these various characteristics are easily modeled, using simple animation programs, on almost any personal computer.

The Positron. The positron can be pictured as structurally similar to the electron, but, as indicated above, the oscillations have an inward moving phase as opposed to the electron's outward moving phase.

The Neutrino. It seems that the neutrino may be the top half of an electron coupled with the bottom half of the positron. Quoting from MLET's author[74]:

> But a strange thing happens to this particle on its way to reality. The phase of the negative north pole still moves outward at the speed of light, and the phase of the positive south pole still moves inward at the speed of light Thus the energy of the particle will be minimized if it moves in the downward direction (opposite to its spin vector) at the speed of light or at least very close to the speed of light.

♣ And what about quarks?

→ Now we're getting into a more speculative arena again.

The various quarks may be electrons and positrons with the addition of *physical* spin. Among other things this is suggested by their mass; spin can create a large local kinetic field whose energy is seen as mass. If the physical spin model is correct, then the various colors and flavors are probably the results of specific spin axis orientations coupled with specific spin magnitudes. And the reason you'd never find them as isolated particles is because physical spin cannot be maintained (is an unstable phenomenon) except in specific multi-particle constructs.

♣ You suggested a source for the strong nuclear force in the MLET basics, but refresh my memory, especially as the force might relate to the quark model

→ If the quark model is correct then it appears that the strong nuclear force might result from a very strong, very localized, gravity field. Remember that an electric field is an oscillating gravity field. If we add physical spin to some combinations of two or three electrons or positrons and place them in very close proximity to each other, it seems that (with the proper physical spin magnitudes and spin axis orientations) the various spin orientations may effect a merging, a smoothing, of the various electric fields and create smoothed density gradient fields — very localized, extremely strong gravity fields.

These intense gravity fields of the three quarks of a neutron or proton bind in the three orthogonal directions inherent in three-dimensional space. The orientation of the three bound quarks then results in a strong three-dimensional gravity field internal to the resulting proton or neutron. The charge of the resulting composite structure is then a function of the combination of the various spins (both intrinsic and physical).

♣ And the quark's supposed ability to move rather freely within certain constraints?

→ This is a little more speculative than I would like as we haven't yet attempted an actual model. But it seems reasonable to assume that smoothed fields (local, very strong gravity fields) overlap each other, and perhaps the particles themselves, in such a way as to only exhibit their maximum strength as they approach certain specific limits.

♣ And the weak nuclear force?

→ The weak nuclear force seems to result from a similar smoothing of magnetic fields into non-oscillating kinetic fields.

→ You really do seem to think that it's possible to reduce all of particle physics to visualizable phenomena.

→ Yes, we do. We've said that from the beginning.

♣ There are many strange effects inherent in quantum theory. Are you suggesting that you can handle them all intuitively?

→ I think we can, yes. The intuitive explanation may not always be immediately obvious, but yes, I believe MLET can ultimately provide them in all cases. To quote the MLET author[75] once again:

> The most significant improvement in understanding diverse quantum phenomena is, I believe, the movement from "particle" to "standing-wave" pictures of physics. This provides for non-locality and probability wave collapse that remains profoundly puzzling from a particle point of view.

Remember that, in order to maintain resonance, the electron's structure must change with increased velocity. Our computer simulations have shown, and it's rather easy to understand, that the structure must change rather dramatically at speeds that are substantial fractions of the speed of light. We're currently in the process of developing computer simulations of very-high-velocity electrons and, even more interesting, orbiting electrons. We think the orbiting model is going to be especially interesting in the light of the Schrodinger equation.

♣ What about the quantum packages themselves? Of light, of gravity, and so forth.

→ Quantum gravity, gravitons, are fictitious artifacts of some current thinking. In general, to give a simplistic, yet accurate answer, all measurements are necessarily made using real matter. An orbiting electron (due to resonance requirements) can only accept or give up energy in discrete amounts. While that certainly does say something significant about the nature of the electron and of the orbit, it doesn't obviously say anything very significant about the nature of electromagnetic waves themselves.

♣ How do you handle such things as Heisenberg's uncertainty principle?

→ I wouldn't try to answer that here even if I had a completely definitive answer readily available. The lab simulations will, I think, provide a powerful intuitive grasp of why current theory is forced to embrace very strange concepts. The uncertainty principle is only one of them.

♣ How about the famous double-slit experiment showing the wave-particle duality of the electron?

→ Again, we'll be moving to the lab tomorrow, and I think you'll be very surprised at how well simple animations can suggest intuitive answers to such questions.

♣ And black holes, singularities, worm holes?

→ This is an area where MLET really changes things. Although the mathematics of the distortions of an elastic solid ether is remarkably similar to the mathematics of Einstein's space-time curvature, where current theory is pretty much restricted to the mathematics alone, MLET has a visualizable physical model from which the mathematics can be derived. And the concrete physical model tells us things that general relativity never could.

In general relativity energy and matter are equated to the extent that all energy creates gravity. But, as is well known, the energy everywhere in the vacuum doesn't cause the expected curvature. As Michio Kaku[76] writes:

> In fact, the amount of energy flooding the vacuum is 10^{100} times larger than the experimentally observed amount. In all of physics, this discrepancy of 10^{100} is unquestionably the largest.

MLET agrees with current theory that all particles of matter are fundamentally little more than bundles of energy — with one significant implication with regard to gravity. By definition, a 'bundle' is bounded — the energy level of the local ether is not homogeneous in the presence of matter — internal displacement creates an external density gradient, a gravity field. Other forms of energy (ether disturbances) are either locally transitory or distributed in a homogeneous manner — there is no associated density gradient field, no gravity field. The amount of energy flooding the vacuum, the amount of energy everywhere within the ether, is therefore not a problem — observation agrees with theory, homogeneous energy distributions never create gravity. We have no need for the ad hoc additional assumptions required by current theory to accommodate observation.

♣ Homogeneous energy distributions never cause curvature? That's certainly a significant difference.

→ It is, and MLET's ability to map the mathematics to a definite physical model demands a rethinking of such concepts as black holes, singularities and wormholes.

♣ I take it you don't think black holes exist?

→ I think rather exotic phenomena may exist as gravity field intensities approach physical limits, but it's extremely unlikely that the name 'black hole' would be appropriate to any sensible model of anything that actually happens.

♣ You still have gravitational waves to deal with.

→ With only two fundamental field types there's no room for (and no need for) a new type of wave. Gravitational waves are gravitokinetic (electromagnetic) waves, and will never be detected by the LIGO type experiments.

- ♣ You're certainly quick with explanation. I still suspect most are rather off base, but they do seem interesting at first blush. But without rather rigorous justifications they can never be very convincing.
- → Again, the nice thing about intuitively visualizable solutions is that intuitively understandable computer simulations, very specific simulations, are possible for every aspect of the theory. Over the next few weeks you'll see some quite rigorous demonstrations of what we believe is going on.
- ♣ And, when do I get to see these simulations?
- → We got the lab computers set up and operational last Wednesday. Our graphics people were new to me so I wasn't sure how soon they'd be ready for us. But they tell me they'll be ready tomorrow, so we'll be using the simulations from here on out. They think they can keep ahead of us.

    The first simulations illustrate the derivation of the Lorentz equations, going from the underlying asymmetry to the observed symmetry. We'll illustrate the Michelson-Morley results and move on to general resonance requirements.

    We'll be modeling individual electrons travelling at various speeds, see the effects of accelerations, attempt to model interactions between electrons and move on from there. I think you'll find what they're doing interesting.

    We'll still require your one hour a day, but I'll probably not spend more than one or two morning sessions a week with you. You'll have a pretty free reign to ask questions, suggest problems needing resolution, whatever. And feel free to wander about the place. As I said, most of the immediate activity is in the graphics lab right now. But feel free to visit with

any of our staff.
- ♣ I'm looking forward to the change of format.
- → Good. Meet me here in the morning and I'll take you to the lab.

— *** —

"So this is my last day in the interrogation cell?" David asked as he came into my office.

"For a while at least," I assured him. "If things go as planned we won't be back here for more than one or two sessions a week at the most, until your last day here. At that point we do want to record your thoughts, impressions, beliefs, those sorts of things, before you leave us. Remember — one hour a day — otherwise come and go as you please."

After Thirty Days
[Jim Price and David Rhodes]
14 June, 1999

→ So how do you feel about your time with us?

♣ Before we get into that, could you answer one question?

→ Sure.

♣ You've committed fifty million dollars to this effort. That's a fortune that could have been spent very effectively in many other areas. What motivated this kind of commitment?

→ The short answer is that I believe in MLET and think it's important that the physics community become aware of the better alternative.

♣ But to the tune of fifty million dollars?

→ One of my concerns has to do with what the physics community has recently been advocating with regard to all science teaching. Some have proposed to emphasize, very early in the student's exposure to any science, that, as Davies put it, "not all that is so can be grasped by the human imagination." Supposedly, this 'don't expect to understand everything science tells you, because no one really does,' would make SRT and quantum concepts less troubling when first encountered by the science student.

I believe that it's very important to the advance of science that these troubling concepts *remain* troubling. One reason for my very large commitment to the Tahoe Project is that I think that this determination to kill any remaining skepticism with regard to SRT could be extremely damaging to the future of physics — if successful, devastating.

♣ Given that you believe as you do, that makes sense. But what gave you this level of confidence in MLET?

→ Before I ever heard of MLET (my Bachelor of Science degree from Stanford was a dual major, physics and electrical engineering) I was troubled by the absence of any significant evidence in support of special relativity.

♣ What do you mean by that?

→ I was one of those who never liked the apparent paradoxes when one went beyond mere ability to predict the results of events that can be measured. So, in my senior year I spent considerable time researching what really took place in the development of theoretical physics during the first decades of the twentieth century. I came to believe, as Steven Mitchell puts it, that, "to accept Einstein's interpretation as proven fact, one must first reject the better explanation." The Lorentz interpretation just made better sense.

♣ But, as I understand it, Lorentz's concepts were never really as complete as SRT. Some concepts were not as fully developed.

→ That's probably true, but it's easy to understand why. SRT was fundamentally an assertion that 'the mathematics, the measured values, are the reality' — questions going beyond that were made meaningless. On the other hand, Lorentz took the position that the measured values were telling us things with regard to an underlying physical reality. In marked contrast to SRT, this required the asking and answering of a lot of very difficult questions. So, while Lorentz made a very solid start, it's true it wasn't as complete as SRT — but again, that was simply because SRT disallowed the most important questions. MLET is, in many ways, little more than a logical extension of what Lorentz began.

To my mind, Einstein grossly over-simplified and so gained an immediate advantage, but at the expense of ultimate confusion. Lorentz took the more conservative, more cautious path, and was pushed aside by those preferring more immediate answers. In the long run the Lorentz approach is not only the safest, but by far the surest and simplest, approach to good understanding.

♣ You believed all that, but instead of challenging the consensus, you simply walked away. Why?

→ It wasn't quite that simple. Frankly, electronics was my first love. But I did make some small effort to make my viewpoint known. However, after voicing my skepticism with regard to SRT to a few of my professors and peers, I quickly became convinced that I'd have to dedicate my entire professional life to theoretical physics to have any chance of convincing anyone, and even then, the likelihood of having significant impact would be virtually nil. So, yes, I walked away. As I'm sure others have also.

♣ But now?

→ That was at the beginning of my career, this is forty years later. Then failure to convince would have been devastating, now I have a committed nucleus that makes the effort fascinating and satisfying — and I have MLET, with a firm prediction that the LIGO observatories will fail to detect gravitational waves.

♣ And if they do?

→ They won't. The massive knowledge base, the known mathematical relationships, the whole of modern physics tells me that MLET is right — the 'ripples in

space-time' that they're looking for simply don't exist.
- ♣ We'll see, won't we?

  Do you expect Julia's book to have much impact?
- → It'd be crazy to expect it to have any immediate impact on the consensus community. I doubt they'll even notice it. And, if it did happen to fall into the hands of any true believers, I doubt that any would get further than the first page or two. That's to be expected- they know we're wrong.
- ♣ Then why bother?
- → You probably know the answers to that. First, we want to be on record before the LIGO observatories are operational. Second, we think there may be a larger than generally thought dissident (or just uneasy) community out there.
- ♣ You may be right. I guess you've answered my question about your motivation..
- → Good. So how do you feel about your time with us?
- ♣ To tell you the truth, I think I may be in trouble. Steve warned me that any inclination to support MLET would kill me professionally. I'm afraid he may be right.
- → You're perfectly free to walk away and not look back. You've met your obligations as far as I'm concerned.
- ♣ Easy to say. Trouble is, I enjoy a good challenge. Especially when I think there's a chance to really learn something, to possibly make a big difference.
- → You're saying?
- ♣ MLET fascinates me. There's certainly some unresolved questions, but unless I find something really intractable, I'm going to have trouble walking away.
- → Initial LIGO results will probably be in within the next

two or three years, so why not forget MLET 'till then? Then decide with much less professional risk.

♣ And if they definitely find what they're looking for right away, I'll be glad I waited. On the other hand, if the results are negative, I'll have missed a lot of the fun.

→ From your perspective there's some risk. What do you see as the potential benefits of staying with us?

♣ If MLET is right it's a really big deal. I wouldn't want to miss out.

→ So?

♣ I think if I had to decide right now, I'd go with MLET.

→ But you wouldn't be completely comfortable with that?

♣ Scared to death.

→ If you're leaning at all toward MLET you must have some reasons. I'd be very interested in those if you wouldn't mind discussing them.

♣ Mansouri and Sexl, Zahar and others make some very, very strong points. And the lab simulations reinforce such Zahar[77] statements as:

> Lorentz was justified in asserting that:
> " . . . the chief difference [is] that Einstein *simply postulates* what we have *deduced* . . . from the fundamentals of the electromagnetic field."
>
> Lorentz explained Michelson's result in a non *ad hoc* way; he was first to discover the transformation laws for the electromagnetic field; he described the way in which the inertia of the electron depends both on its energy and on its velocity; and he explained the invariance of c. Thus Lorentz's continued adherence to his own programme after 1905 was completely rational.
>
> Lorentz's theory was eminently intelligible whereas

Einstein's involved a major revision of our most basic notions of space and time.

I think a very good case has been made in favor of MLET. And with regard to general relativity and quantum theory also.

→ Julia, I think, would find it hard to resist an 'I told you so.'

♣ I'll bet.

But I am in a quandary — a bit afraid to make any decision. On the one hand, possibly unprecedented opportunity, on the other potential loss of professional credibility.

→ Maybe you'll still find the intractable problem that will let you reject MLET and comfortably return to the consensus community.

♣ Looking at my personal conclusions and how faithfully they reflect Mitchell's original claims, I sometimes wonder if I haven't been the subject of a very successful brain-washing experiment.

→ I understand. You have a lot to lose.

We've suspected you might be leaning our way over the last few days and we do want to be careful not to force a decision. If you want, I'd suggest that you spend a month (longer if you need) in a more neutral environment before you make any serious commitment. I'll make sure you won't suffer financially, regardless of your final decision.

♣ I appreciate that. I probably should take you up on it, but at this point I really don't want to leave the lab. The advanced simulations are a little too intriguing to easily walk away from.

→ That being the case, how about a compromise? Everyone who's worked with you has appreciated your comments and suggestions. So stay on as long as you like, leave when you want.

♣ Something tells me you understand my current situation pretty well.

→ We did a lot of research before I made the initial offer. Among other things I talked to some of your friends at both Stanford and CERN.

♣ So you knew I wasn't happy with where my work seemed to be taking me and had already asked about an extended leave?

→ Yes, I was aware of that.

♣ I'll admit that ever since the Superconducting Super Collider effort was cancelled, I've wondered about opportunities in theoretical physics. I stayed with it, even though I've seriously considered switching to micro-biology more than once. Something tells me that the possibility of my leaving physics was a factor in your selecting me.

→ It was. Even our most limited goal was to create doubt with respect to the absolute certainty of SRT. Everyone seemed to agree that if we were at all successful, we probably would kill any chance of your enjoying any further work at CERN. Mitchell didn't think we had much chance, but I thought we might.

So I liked the fact that, from what your friends told me, loss of credibility within the community might not be as devastating to you as it would to others.

♣ That's probably true, but loss of credibility is always painful. And to some extent it would follow me wherever I go.

→ So you do have a decision to make. We'd certainly welcome your staying on, on a day-to-day basis if you prefer, for as long as you'd like. I would, though, like to suggest some additional options.

♣ Such as?

→ As I'm sure Julia explained to you when you first arrived, we're focusing on three goals: First, rigorously evaluate and define the clear implications of MLET in every area of physics. Secondly, we'd like to give other physicists an opportunity to go through a short program (patterned somewhat on our experience with you, minimizing our demands on their time and maximizing their recreational opportunities, entirely at our expense) introducing them to MLET. As you've seen, MLET involves almost no change to the consensus mathematics so a pretty complete overview is possible in just an hour or so a day over a few weeks.

♣ Do to them what you've done to me?

→ Use what we've learned, yes.

Thirdly, I want to establish a credible graduate level program, with classes in both MLET and current theory — requiring understanding of both, but contrasting the two. In fact, I anticipate that most of the funding will go to establishing a research/graduate level institute focused on MLET. The research part would be a direct outgrowth of our current effort in the lab. But I want to be sure that all students will also be fairly well grounded in the leading edge of current theory in order to gain and retain credibility.

♣ Where do I fit in?

→ I'd like to transition the theoretical development, including our laboratory effort to a credible research program open to qualified students by the fall of 2001. To do that, we have to be able to answer many questions in the affirmative. Can we interest (and get up to speed) at least two or three people with appropriate credentials to serve as professors/mentors/counselors? Ph.Ds with good credentials? Can we develop the laboratory tools and teaching aids that will be needed? Where, how, do we get qualified and interested students?

You get the picture.

I can make sure that physical resources — facilities, computers, whatever — will not be a problem.

♣ Very ambitious if you expect to do a credible job.

→ Everyone currently on board will be concentrating on path one and two goals for the next six months to a year. The research effort should easily be transitioned to the long term effort. If you come with us, I'd like for you to join Steve, Julia and me in establishing and maintaining the focus needed to meet the fall 2001 goal.

♣ I think I'd prefer concentrating on defining, explicitly, the scope and implications of MLET, with development of sophisticated computer simulations — general MLET related research.

→ You have strong interests. Good. You'd certainly be able to spend most of your time in those areas — credible courses will require exactly what you clearly hope to provide. I promise you wouldn't need to give us more than four hours or so a week toward the establishment of the research/teaching program. In any case, don't think you have to make any decision immediately.

♣ I'm intrigued. But, if it's really all right with you, and as long as I'm free to continue my work in the lab, I would like to delay any firm decisions for a month or so. I need to do some serious thinking.

→ That's fine.

There's one other subject I'd like to discuss with you

this morning, though. Would you mind staying a few more minutes?

♣ Not at all.

→ You know that Sarah, Paul and Tim just recently got their bachelor's degrees in physics. Have you wondered how they came to work here?

♣ I've been impressed with their understanding of how the computer simulations should be structured to reflect the MLET perspective, and I'll admit, I did initially wonder how such intelligent and knowledgeable people had ended up with you people.

→ I tried something with them that I'd like to repeat with others at their level. As I told you earlier, I graduated with a double major, physics and electrical engineering. I didn't continue in physics largely because I didn't really like current theory.

When we wanted to expand our effort in exploring and explicitly documenting the MLET implications, I remembered my own distaste for the prevailing consensus and suspected that there were probably other physics majors who were less than enthralled by where physics theory seems to be going.

As you can imagine, with some fifty thousand employees, I've maintained contacts at many of the major universities. So I went looking for bright young new graduates, very bright young people that the physics departments were afraid of losing. I was interested in those skeptical of theory or opportunity, or both. I was pleased to find that there were a fair number of students to choose from.

♣ That surprises me. But maybe it shouldn't. I do remember that some of my undergraduate friends had trouble accepting, as one of my professors put it,

'Nature as She is.'

→ It was interviewing these students that convinced me that we needed a viable research/teaching facility to let these students contribute. I expect them to be our surest path to eventual credibility — surest, but definitely long range.

You were a different matter. I became aware that there were some within the community that were troubled in various ways with regard to both the status and direction of current thinking. It was more widespread than I had realized, but clearly wasn't something we were in any position to take immediate advantage of. So I was pleased when Julia proposed that as a first step, we concentrate on a single, credible physicist, and see if we could have an impact.

♣ It's still rather shocking to me that you did have an impact.

→ Do you think we could have a similar impact on others like you?

♣ A few weeks ago I would've laughed in your face at the mere suggestion. But now? Yes, I'm sure there are many who might be swayed just as I was.

→ How soon do you think we might have an impact on mainstream thinking?

♣ I'd have to say that on that point, I'm pretty much in agreement with Steve. The consensus will probably hold for decades.

→ That long?

♣ When I first became convinced that MLET was probably right, I thought that, with the right approach, acceptance might come fairly rapidly — maybe within five to ten years. But after closer examination of current attitudes (what the community really believes with respect to the possibility of Einstein being wrong) I began to appreciate that Dingle was probably right — they've let their belief in SRT escape the bounds of reason, have lost all ability to allow questioning. This is seen in much of the current literature in the form of the complete inability to understand that the facts, the supposed proofs of SRT, do not require SRT's interpretation. So no matter how much better MLET is in explaining observation, it is highly unlikely that they'll ever seriously examine the claims. Something Hawking[78] wrote illustrates the problem you'll have in reaching the community's controlling elite:

> There is a subspecies called philosophers of science who ought to be better equipped. But many of them are failed physicists who found it too hard to invent new theories and so took to writing about the philosophy of physics instead. They are still arguing about the scientific theories of the early years of this century, like relativity and quantum mechanics. They are not in touch with the present frontier of physics.

MLET's advocates will be seen as still arguing about settled beliefs — simply out of touch with the reality revealed by modern physics. So I'd expect that, no matter how strong the evidence, the current community leaders will probably have to die off before the consensus can change to the point where the leading universities would ever touch any of your concepts.

→ That bad?

♣ I expect so. It took me a while to really appreciate one point Steve made. When a Richard Feynman can find absurdity 'delightful' it seems highly unlikely that superior explanation will be sufficient to gain a hearing — there's obviously a very strong attachment to a Nature that exceeds all rational bounds. Again, once one embraces absurdity as absolute fact, one has so compromised intellectual integrity that there is almost no likelihood that one might be swayed by rational argument — and sensible explanation becomes a negative in itself. So much so, that even superior ability to predict things that can be measured, when coupled with a claim of superior explanation, will be extremely difficult to accept as convincing.

→ Then how would you proceed?

♣ I think your approach could hardly be improved on. Young graduates who aren't completely happy with current theory and Ph.Ds who question their chances of contributing to leading edge physics are the natural targets of opportunity. I think there may be a surprisingly large number who would jump at the chance to make a difference.

→ In spite of the fact that they'd be 'outside the pale,' so to speak — perhaps for decades?

♣ Once they're convinced you're right, I think that would only add to the appeal. They'd see themselves as the new revolutionaries. And, as long as the current consensus holds there'd be almost no competition from mainstream theorists — a very unusual opportunity for an individual to make a significant contribution.

→ I appreciate your comments. You make the same arguments Julia has insisted on from the start, and she convinced me.

♣ She's something else. I'm not sure I wouldn't have walked away during the first week, but for her. Where'd you find her?

→ Ask her — it's an interesting story.

I'll be spending all summer here at the lake, so I'll be available anytime, if you have any other questions.

— *** —

As I put away the one last tape, David entered my office.

"So tell me an interesting story," he said.

"I'm leaving for Palo Alto in about two hours," I replied. "So not today. If you're still around in a month or so- who knows."

"Before you go," David said. "I'd like to ask you to do something for me."

"I could try. What do you have in mind?"

"I know you're planning to publish our conversations in book form to help publicize the MLET prediction with regard to LIGO. I have no problem with that and I've read your account of all of the conversations to date. But I have a favor to ask. There were a lot of references to what others have written and said with regard to physics theory. I could flip back and forth between the full text and the typical reference list and maybe get what I want, but I'd like to see the full quotation, who introduced it (me, you or Steve), a comment as to why it was introduced in the particular context, and the author and source of the quoted material, all together, for each quotation."

"That's a little unconventional. I could try it and see whether you think it worthwhile."

"I'm sure I'd think it worthwhile. I think the average reader would also. At the very least it would provide an opportunity to review important points in a slightly different context."

"OK. I can do it. If you like, we'll include a notes and references section in the form you've suggested. I'll be back in three days and should have it ready for your review within a week or so."

Postscript
26 July, 1999

David's still with us.

I had planned to include some supporting documents in an appendix, but changed my mind after spending some time in the graphics lab. Instead, I've decided to publish all of the supporting documents later under a separate cover to allow us to include more graphics, and, hopefully, computer-generated simulations on a CD-ROM as part of the supporting material. The graphics add so much impact to the arguments that it seems well worth some months' delay to be able to include them.

I could have delayed publishing anything, but decided that it was important, given the current status of the LIGO observatories, to get the gravitational wave prediction, along with some basic theory, published as soon as possible.

So, for now, here's the conversations. (I have included a "Notes and Quotes" section in the form David suggested.)

Watch for the supporting documents and computer simulations. I think you'd enjoy them.

Julia Clark

Notes and Quotes

This section is structured as David Rhodes suggested. The first line of each entry is composed of the reference number, the author of the excerpted material, and the person who introduced it to the conversation. This is followed by the actual quotation (in smaller type and indented both left and right) and notes or comments (in normal size and format). In some cases the comments are nothing more than the context within the conversation. The notes, comments and quotation are then followed by the source in a more conventional reference format. When a particular quote was used more than once, it is listed here each time it was used.

1. Albert Einstein   Steven Mitchell

    It would be a great advantage if we could succeed in comprehending the gravitational field and the electromagnetic field together as one unified confirmation.

    Einstein introduced the intractable disparate domain boundaries, then spent the last decades of his life trying to breach them. He failed, as have all others. LIGO's results may finally force the consensus community to allow essential questions.

    Einstein, Albert (1983) "Ether and The Theory of Relativity," (An address delivered on May 5th, 1920, in the University of Leyden) *Sidelights on Relativity*, Dover Publications, Inc, New York, p 22.

2. George Gamow   Steven Mitchell

    Albert Einstein became the ruler of modern physics by cutting the ethereal knot with the sharpness of his logic, and throwing the twisted pieces of world ether out of the window of the temple of physical science.

This quote is not untypical of the current disparagement of the concept of an ether. This makes it difficult for any ether theory to get a hearing, especially since any such theory must disagree with SRT.

Gamow, George, (1961) *Biography of Physics*, Harper & Brothers, New York, p 159

3. Albert Einstein   Steven Mitchell
   Since the special theory of relativity revealed the physical equivalence of all inertial systems [his absolute symmetry at all levels] it proved the untenability of the hypothesis of an ether at rest. It was therefore necessary to renounce the idea that the electromagnetic field is to be regarded as a state of a material carrier. The field thus becomes an irreducible element of physical description . . .

Here we find the true source of the domain boundary problem. Absolute physical equivalence of all inertial frames (absolute symmetry at all levels) makes it forever impossible to develop a single convergent theory encompassing both electromagnetic and gravitational phenomena.

Einstein, Albert, (1954) "Relativity and the Problem of Space," R.W. Lawson Trans. *Relativity, The Special and the General Theory*, Methuen and Company, Ltd., London

4. Paul Davies   David Rhodes
   If relativity were wrong, our detailed understanding of subatomic physics would collapse. . . From quarks to quasars, scientists would no longer be able to understand the basis of their own immense knowledge.

It is almost universally believed that modern physics is firmly based on Einstein's relativity, making it impossible to reject his theory and still retain a viable science. This belief is wrong. Identical data and virtually the same mathematics better support an ether theory. The Modified Lorentz Ether Theory allows derivation of the current mathematics from a sensible physical model, so that, far from any loss of understanding, with the transition to the better theory the "immense knowledge" is really *understood* for the first time ever.

Davies, Paul. (1980) "Why Pick on Einstein?," *New Scientist,* Aug. 7., 1980

5. Stephen Hawking   Steven Mitchell
    I take the ... viewpoint that a physical theory is just
    a mathematical model and that it is meaningless to
    ask whether it corresponds to reality.

Current theory makes any physical reality meaningless — Hawking accepts that.

Hawking, Stephen W. and Roger Penrose (1996) *The Nature of Space and Time*, Princeton University Press, Princeton.

6. Albert Einstein   Steven Mitchell
    For the theoretician such an asymmetry in the
    theoretical structure, with no corresponding
    asymmetry in the system of experience is
    intolerable.

With this rejection of the asymmetry, Einstein justified rejection of good explanation and so opened the way to acceptance of the absurd.

Einstein, Albert (1983) "Ether and The Theory of Relativity," (An address delivered on May 5th, 1920, in the University of Leyden) *Sidelights on Relativity*, Dover Publications, Inc, New York, p 12.

7. Leon Lederman   David Rhodes
    The tragedy in all this is not the sloppy pseudoscience writers, not the Wichita insurance salesman who knows exactly where Einstein went wrong and publishes his own book on it. It is the damage done to the gullible and science-illiterate general public, which can so easily be duped.

    The scientific community bears a considerable responsibility to protect the general public from shoddy and misleading pseudo-science. That's a given. Frustrating as the process may be, the dissenter must bear the burden of proof.

    Lederman, Leon with Dick Teresi (1993) *The God Particle*, Houghton Mifflin Company, New York, p 193

8. Paul Davies   Steven Mitchell
    Using gravity-wave detectors as 'gravity-telescopes' is on the horizon. With such a facility we could 'see' into the dense hearts of quasars and neutron stars, probe to the very edges of black holes and maybe eventually listen to the rumble of the primordial big bang itself.

    The failure of the LIGO observatories to detect gravitational waves will be devastating to the expectations of the consensus community.

    Davies, Paul. (1980) *The Search for Gravity Waves*, Cambridge University Press, Cambridge

9. Paul Davies   Steven Mitchell
    Most editors of science magazines and journals make special provision for coping with the huge influx of papers and letters, many bearing private addresses in California, purporting to disprove or improve Albert Einstein's monumental work on the theory of relativity.

Davies seems to ascribe the 'huge influx' to ignorance and inability to accept the non-intuitive nature of Einstein's special relativity theory. There is a lot of ignorance, but I would argue that the rejection of SRT's implications has more to do with the lack of consistency, coherence and convergence in current theory than with any innate inability to accept the non-intuitive. Any reasonably intelligent person must suspect that something is seriously wrong, and very many of these people have correctly guessed that the problem lies at the very foundation of the current consensus — with SRT itself.

Davies, Paul (1980) "Why Pick on Einstein?," *New Scientist,* Aug. 7, 1980.

10. Leon Lederman David Rhodes

> . . . what is rarely understood by the lay public is how ready, how eager, how desperately the collective science community in a given discipline welcomes the intellectual iconoclast — if he or she has the goods.

Lederman argues that, given 'the goods,' the consensus community will readily embrace iconoclastic concepts — we disagree. The collective science community is, on the whole, much too sensible to allow acceptance of iconoclastic concepts, given that any supposed 'goods' must, by definition, contradict settled belief — the consensus community's truth. Acceptable 'goods' can never contradict known truth.

Lederman, Leon with Dick Teresi (1993) *The God Particle,* Houghton Mifflin Company, New York, p 193.

The following references (11-16) were introduced by Steven Mitchell in support of his argument that the current consensus is much too strong to allow questioning of the absolute truth of

SRT, no matter the strength of contrary evidence

11. Carl Lanczos   Steven Mitchell

> Nobody intends to diminish the merits of other great men of science, but there was something in Einstein's mental make-up which distinguished him as a personality without peers. He wrote his name in the annals of science with indelible ink which will not fade as long as men live on earth. There is a finality about his discoveries which cannot be shaken. Theories come, theories go. Einstein did more than formulate theories. He listened with supreme devotion to the silent voices of the universe and wrote down their message with unfailing certainty.
>
> What was so astonishing in his manner of thinking was that he could discover the underlying principle of a physical situation, undeceived by the details, and penetrate straight down to the very core of the problem. Thus he was never deceived by appearances and his findings had to be acknowledged as irrefutable.

Carl Lanczos, in a lecture titled "The Greatness of Albert Einstein," delivered at the University of Michigan in the spring of 1962 as part of a lecture series titled "The Place of Albert Einstein in the History of Physics," From *Albert Einstein's Theory of General Relativity: 60 years of its influence on man and the universe,* Gerald Tauber Ed. Crown Publishers, Inc., New York, 1979, p 16

12. Nigel Calder   Steven Mitchell

> Einstein's theories are the bedrock. . . . It *is* Einstein's universe.

Calder, Nigel (1979) *Einstein's Universe*, The Viking Press, New York, p 3

13. Paul Davies and John Gribbin   Steven Mitchell

> All of the implications of special relativity . . . have

been confirmed by direct experiments. There are still people who believe that it is all "just a theory" . . . but *they are wrong.*

Davies, Paul and John Gribbin (1992) *The Matter Myth*, Simon & Schuster, New York.

14. John Gribbin   Steven Mitchell

Gribbin claims that Einstein's concepts are

> . . . fully accepted by all except a tiny minority equivalent to the flat-Earthers, who still don't believe the Earth is round.

Gribbin, John (1979) *Time-Warps*, Delacorte Press, New York, p 69.

15. Isaac Asimov   Steven Mitchell

> No physicist who is even marginally sane doubts the validity of special relativity.

As with most SRT advocates, Asimov clearly assumes that no alternative theory consistent with supposed SRT proofs is possible. In this he is profoundly wrong — the facts are not in question — it's simply that the supposed proofs better support the sensible alternative.

Asimov, Isaac (1993) "The Two Masses," *The World Treasury of Physics, Astronomy and Mathematics*, Timothy Ferris Ed., Little, Brown and Company, Boston, p 186. (Excerpted from Asimov's book, *The Subatomic Monster*, published in 1985)

16. Clifford Will   Steven Mitchell

> Special relativity is so much a part not only of physics but of everyday life, that it is no longer appropriate to view it as the special "theory" of relativity. *It is a fact* . . .

Will, Clifford M. (1986) *Was Einstein Right?*, Basic Books Inc., New York.

17. Paul Davies   Steven Mitchell
> In his later years, Dingle began seriously to doubt Einstein's concept of time. He had little difficulty persuading a motley group of followers of the absurdity of relative time . . . the mood of dissent he championed lives on, widespread and festering. I wonder why? Einstein must have touched a raw nerve.

Davies is typical in his contempt for any who might question SRT. That's understandable, he knows SRT is absolute fact, not theory. Davies seems blind by choice to the possibility of better explanation. Understandable, when one is absolutely sure — when theory has been allowed to become fact.

Davies, Paul (1995) *About Time: Einstein's Unfinished Revolution*, Simon & Schuster, New York, p 55.

18. Herbert Dingle   Steven Mitchell
> It is simply that physicists have, unawares, allowed their trust in special relativity to escape the control of reason and become a blind slavery to dogma . . .

Dingle, ridicules as 'blind slavery to dogma' the true believer's refusal to question. And given SRT's lack of any convincing empirical base, it's hard to fault Dingle — it's very hard to understand where the inability to question comes from.

Dingle, Herbert (1976) in a letter to Lord Todd, President of the Royal Society, quoted in The Relativity Question, by Ian McCausland, Dept. of Electrical Engineering, University of Toronto, Toronto, Canada, p 32.

19. John D. Barrow   Steven Mitchell
> . . . scientists . . . receive a large amount of mail from misguided members of the public announcing the discovery of their new "Theory of the Universe"

> (the author has received two during the last week alone). . . . They aim to show how Einstein was wrong in some way and . . . have an obvious psychological motivation. Einstein is perceived as the twentieth-century scientist *par excellence*, and hence it is fondly imagined that, by catching him out on some point, the new author would be hailed as the new scientific messiah, greater than Einstein.

So questioning Einstein in any nontrivial way has 'an obvious psychological motivation.' No credible physicist can even imagine any valid reason to set aside revealed truth in favor of good explanation.

Barrow, John D. (1991) *Theories of Everything*, Clarendon Press, Oxford, p 87.

20. Stephen Hawking　　　　　　　　Steven Mitchell
> Any physical theory is provisional, in the sense that it is only a hypothesis: you can never prove it.

Maybe. But all seem to agree- there's at least one clear exception. While the word 'theory' is sometimes retained (for obvious reasons, some prefer just 'special relativity') special relativity theory (SRT) *is considered absolute fact*. Hawking's failure (in this case) to appropriately qualify a general principle can be quite confusing to the culturally deprived — Davies would surely object strenuously to Hawking's oversight.

Hawking, Stephen W. (1988) *A Brief History of Time*, Bantam Books, New York, p 10.

21. Richard Feynman　　　　　　　　Steven Mitchell
> Many physical pictures can give the same equations.

And just so, 'the same equations may allow many physical pictures.' Not only do the SRT equations allow a different physical picture, they were *first derived* from a radically different physical picture. Both the data and the mathematics supposed to require SRT, better support MLET's physical picture, the picture that gave us the original derivation..

Gleick, James (1992) *Genius: The Life and Science of Richard Feynman*, Pantheon Books, New York, p 326.

22. Nigel Calder   Steven Mitchell
   This is Einstein's Universe.

Einstein clearly intended to establish his SRT absolute, his absolute symmetry, the physical equivalence of all inertial frames, as a universal, over-arching principle. With his success his 'perfect symmetry' absolute has become the foundation of the prevailing worldview of twentieth-century physics. Few really appreciate that the fundamental concepts underlying this worldview, the concepts that have led to an absurd Nature, really are absurd.

Calder, Nigel (1979) *Einstein's Universe*, The Viking Press, New York.

23. Albert Einstein   Julia Clark
   For the theoretician such an asymmetry in the theoretical structure, with no corresponding asymmetry in the system of experience, is intolerable.

Einstein just couldn't swallow the gnat. In spite of the fact that no experimental data and no natural law required it, Einstein just couldn't allow a Nature that might ever fail to reveal some truth directly to man's 'system of experience.' As a man, that possibility offended him.

Einstein, Albert (1983) "Ether and The Theory of Relativity," (An address delivered on May 5th, 1920, in the University of Leyden) *Sidelights on Relativity*, Dover Publications, Inc, New York, p 12.

24. Richard Feynman   Julia Clark

The more you see how strangely Nature behaves, the harder it is to make a model that explains how even the simplest phenomena actually work. So theoretical physics has given up on that.

Einstein's inability to swallow the gnat, his inability to accept an underlying asymmetry, has had a drastic impact on physical theory — perhaps at no other time in recorded history has 'giving up' been so universally preferred to questioning assumptions.

Feynman, Richard (1985) *QED*, Princeton University Press, Princeton, p 82.

25. Leon Lederman   Julia Clark

All we can ask of a theory is to predict the results of events that can be measured.

Again, disallowance of the underlying asymmetry has done immeasurable damage — has forced the tolerance of disparate domains with the consequent glaring lack of anything approaching universal convergence. The modern physicist clearly understands that ability to predict the results of events that can be measured is about all we have left. And, if Einstein were right, it really would be all we could ever ask.

Lederman, Leon with Dick Teresi (1993) *The God Particle*, Houghton Mifflin Company, New York, p 175.

26. Stephen Hawking   Julia Clark

I take the . . . viewpoint that a physical theory is just a mathematical model and that it is meaningless to

> ask whether it corresponds to reality. All that one can ask is that its predictions should be in agreement with observation.

Hawking well understands the implications of swallowing Einstein's camel. Hold to Einstein's perfect symmetry and you're left with no possible sensible physical reality.

Hawking, Stephen W. and Roger Penrose (1996) *The Nature of Space and Time*, Princeton University Press, Princeton.

27. Paul Davies and John Gribbin   Julia Clark
> I believe that the reality exposed by modern physics is fundamentally alien to the human mind, and defies all power of direct visualization....

And
> The realization that not everything that is so in the world can be grasped by the human imagination is tremendously liberating...

Again, swallow the camel of absolute equivalence of inertial frames, and one finds the rejection of consistency, convergence and cohesiveness in favor of a dumbing-down of acceptable theory to a mere ability to predict, 'tremendously liberating.'

Davies, Paul and John Gribbin (1992) *The Matter Myth*, Simon & Schuster, New York, p 110.

28. Kip Thorne   Julia Clark
> What is the real, genuine truth? Is space-time really flat, as the above paragraphs suggest, or is it really curved? To a physicist like me this is an uninteresting question because it has no physical consequences. Both viewpoints, curved space-time and flat, give precisely the same predictions for any measurement performed with perfect rulers and clocks, and also (it turns out) the same predictions

> for any measurements performed with any kind of physical apparatus whatsoever. . . . they disagree as to whether that measured distance is the "real" distance, but such disagreement is a matter of philosophy, not physics. . . . Which viewpoint tells the "real truth" is irrelevant.

Given mutually exclusive explanations, 'which . . . tells the real truth' is irrelevant? This is a modern physics molded by Einstein and dictated by a lazy, 'the best we can ever do', 'all we can ever ask', 'hold to the dogma at all costs', dumbing down of acceptable theory.

Thorne, Kip S. (1994) *Black Holes & Time Warps*, W. W. Norton & Company, New York, p 400.

29. Albert Einstein   Steven Mitchell

> For the theoretician such an asymmetry in the theoretical structure, with no corresponding asymmetry in the system of experience, is intolerable.

One must suspect that good science should require that man must ultimately defer to Nature with respect to what concepts should be allowed — the young Einstein had no such inclination. Positivism insisted that going beyond direct experience was intolerable (placed Nature in a straight-jacket of man's design) and Einstein pronounced it to be so. Having acknowledged that Lorentz, with his underlying asymmetry, had succeeded in showing that measured values (such as the Michelson-Morley results) were not inconsistent with an ether at rest, he proceeded to find good explanation intolerable and so made absurdity sacrosanct.

Einstein, Albert (1983) "Ether and The Theory of Relativity," (An address delivered on May 5th, 1920, in the University of Leyden) *Sidelights on Relativity*, Dover Publications, Inc, New York, p 12.

30. Max Born   Steven Mitchell

With regard to a concept that "was introduced to explain." Born writes:

> Sound epistemological criticism refuses to accept such made-to-order hypothesis.

Born insists that one can never justify violating certain 'epistemological' principles just to get good explanation, just to provide consistency, coherence, convergence, clarity, a rigorous mathematics and a sensible physical reality. Such a violation, is, as Einstein claimed, 'intolerable' — and Nature cannot be allowed to object.

Born, Max (1962) *Einstein's Theory of Relativity*, Dover Publications, New York, p 310.

31. Albert Einstein   Steven Mitchell

> It is often, perhaps even always, possible to adhere to a general theoretical foundation by securing the adaptation of the theory to the facts by means of artificial additional assumptions.

Here was Einstein at his best. In the light of doubts expressed to Solvine (Note 38), one might suspect that if he were alive today, he would see much of current theory as adhering to his 'general theoretical foundation by securing the adaptation of the theory to the facts by means of artificial additional assumptions.'

Einstein, Albert (1993) "Autobiographic Notes," *The World Treasury of Physics, Astronomy and Mathematics*, Timothy Ferris Ed., Little, Brown and Company, Boston, p 585.

32. Richard Feynman   Steven Mitchell

> Many physical pictures can give the same equations.

Mathematics alone cannot give definitive answers, yet some can't allow the obvious alternative to SRT — no matter that it works much better.

Gleick, James (1992) *Genius: The Life and Science of Richard Feynman*, Pantheon Books, New York, p 326.

33. James Gleick   Steven Mitchell

    He had a set of practical tests, heuristics, that he applied when reaching a judgment about a new idea in physics: . . . for example, did it explain something unrelated to the original problem? He would challenge a young theorist: *What can you explain that you didn't set out to explain?*

Such unexpected benefits are characteristic of convergence. This is why very modest original goals led to the full realization of MLET — some slightly different basic assumptions demonstrated universal convergence within an intuitively understandable physical reality.

Gleick, James (1992) *Genius: The Life and Science of Richard Feynman*, Pantheon Books, New York, p 369.

34. Richard Feynman   David Rhodes

    I'm going to describe to you how Nature is — and if you don't like it, that's going to get in the way of your understanding it. It's a problem that physicists have learned to deal with: They've learned to realize that whether they like a theory or they don't like a theory is *not* the essential question. Rather, it is whether or not the theory gives predictions that agree with experiment. It is not a question of whether a theory is philosophically delightful, or easy to understand, or perfectly reasonable from the point of view of common sense. . . . So I hope you can accept Nature as She is — absurd.

No matter how much you might like 'philosophically delightful' explanation, a modern physics constrained by Einstein's absolute symmetry requires an absurd Nature. And, given only that SRT is absolute fact, Feynman must encourage his audience to accept Her as Einstein required — absurd. And once absurdity is wholeheartedly embraced one is clearly no longer susceptible to the normal influence of rational objection.

Feynman, Richard (1985) *QED*, Princeton University Press, Princeton, p 10.

35. Stephen Hawking   Steven Mitchell
    I take the positivists' viewpoint that a physical theory is just a mathematical model and that it is meaningless to ask whether it corresponds to reality. All that one can ask is that its predictions should be in agreement with observation.

Hawking's 'all that one can ask' is an almost universal belief within the consensus community — but few so frankly acknowledge the positivistic base.

Hawking, Stephen W. and Roger Penrose (1996) *The Nature of Space and Time*, Princeton University Press, Princeton.

36. Leon Lederman   Steven Mitchell
    All we can ask of a theory is to predict the results of events that can be measured. This sounds like an obvious point, but forgetting it leads to the so-called paradoxes that popular writers without culture are fond of exploiting.

Lederman, in confirming Hawking's perspective, notes how well it protects theory from reasonable criticism — any possible objection must, at least implicitly, demand more of theory than mere ability to predict. His obvious point is that no matter how much we may prefer sensible understanding, Einstein has robbed theory of any fundamental requirement for consistency, coherence and convergence — so live with it — you really must, since SRT is known to be absolute fact.

Lederman, Leon with Dick Teresi (1993) *The God Particle*, Houghton Mifflin Company, New York, p 175.

37. Steven Mitchell   Stephen Mitchell
    To accept Einstein's interpretation as proven fact, one must first reject the better explanation.

Even in the early decades of this century, Lorentz and others had already provided a solid base for a much better explanation of experimental data than SRT. To make SRT even remotely acceptable, it was necessary to show that the better explanation was unacceptable — and only positivism could provide any semblance of an excuse to set aside the better theory — and, just so, Einstein used positivistic principle to declare the better theory intolerable.

38. Banesh Hoffmann (of Einstein)   Steven Mitchell
    To Solvine, who had written congratulating him on his seventieth birthday, he wrote in reply on 28 March 1949, saying in part: "You imagine that I look back on my life's work with calm satisfaction. But from nearby it looks quite different. There is not a single concept of which I am convinced it will stand firm, . . ."

To his credit, Einstein didn't see himself in the same light as did his peers. On the other hand, unable to fully accept the likely implications of many years of failure, he remained, to the end, committed to the futile struggle to make relativity theory a sensible theory.

Hoffmann, Banesh (1972) *Albert Einstein: Creator and Rebel*, The Viking Press, New York, p 257.

39. Michael Levine   Steven Mitchell
    Some ideas are so stupid that only intellectuals could believe them.

Strongly held belief, presented to the student as fact, and constantly reaffirmed over many years, does sometimes destroy critical inclinations within the elite community.

Levine, Michael (1995) *Lessons at the Halfway Point*, Celestial Arts, Berkeley, California, p 62.

40. Isaac Asimov   Steven Mitchell
    A clock in motion, he said, keeps time more slowly than a stationary one. In fact, all phenomena that change with time change more slowly when moving than when at rest, which is the same as saying that time itself is slowed.

No clock ever does more than measure rate of change. And no one can doubt the observed fact that, with increased velocity, the measured rate of change of all things that change with time, will be slower. Yet, only the confirmed positivist should ever claim that this 'is the same as saying that time itself is slowed.' Einstein required it, but no conceivable experiment can ever prove it.

Asimov, Isaac (1984) *Asimov's New Guide to Science*, Basic Books, New York, p 390.

41. Richard Feynman   Steven Mitchell
    The more you see how strangely Nature behaves, the harder it is to make a model that explains how even the simplest phenomena actually work. So theoretical physics has given up on that.

Welcome to 'Einstein's Universe.' The enthralled, unquestioning consensus community prefers simply giving up to allowing questioning of Einstein's absolute.

Feynman, Richard (1985) *QED*, Princeton University Press, Princeton, p 82.

42. Max Born   Steven Mitchell

> A concept refers to a physical reality only when there is something ascertainable by measurement corresponding to it in the world of phenomena. This is not the place to enter into a discussion on the philosophic concept of reality; it is at least certain that the criterion of reality just given corresponds fully with the way the word "reality" is used in the physical sciences. Every concept that does not satisfy it has gradually been eliminated from the structure of physics.

And some still claim that positivism is not an essential part of modern theory. Born immediately follows the above with:

> We see at once that in this sense a "fixed spot" in Newton's absolute space has no (physical) reality.

One gets a little tired of Born's 'the way the word is used in the physical sciences' and 'in this sense' to imply an acknowledgement of less than absolute surety, when all the time, he really, strongly, insists that: 'If we can't directly detect it, it ain't real.' Period. Exclamation point! Science will absolutely not tolerate any other interpretation. Although this claim is laughingly absurd in the light of leading edge cosmological theories, it is rigorously true with respect to concepts that might call into question Einstein's absolute symmetry concepts — concepts such as Lorentz's ether.

Born, Max (1962) *Einstein's Theory of Relativity*, Dover Publications, New York, pp 69-70.

43. Steven Weinberg   Steven Mitchell

> Positivism helped to free Einstein from the notion that there is an absolute sense to a statement that two events are simultaneous; he found that no measurement could provide a criterion for simultaneity that would give the same results for all observers. This concern with what can actually be observed is the essence of positivism. Einstein acknowledged his debt to Mach.

And Weinberg continues;

> Despite its value to Einstein and Heisenberg, positivism has done as much harm as good.

In the hands of Einstein and Heisenberg positivism has done its most lasting harm. One would be hard pressed indeed to find any good attributable to positivism.

Weinberg, Steven (1992) *Dreams of a Final Theory*, Pantheon Books New York, pp 175-176.

44. Isaac Newton   Steven Mitchell

> It may be, that there is no such thing as equable motion, whereby time may be accurately measured. All motions may be accelerated and retarded, but the true, or equable progress, of absolute time is liable to no change. The duration remains the same . . . whether the motions are swift or slow, or none at all . . .

Born rejects Newton's reasoning on positivistic grounds. We reject Born's grounds, and find Newton's argument as strong as ever.

Born, Max (1962) (quoting Isaac Newton, 1729 translation by Andrew Motte) *Einstein's Theory of Relativity*, Dover Publications, New York, p 57.)

45. Steven Weinberg David Rhodes
(Comments by Mitchell)
> He found that no measurement could provide a criterion for simultaneity that would give the same results for all observers.

Einstein was right in his finding. But to treat inability to measure as proof that absolute simultaneity is an impossible concept must be seen as ridiculous to any but the most rabid positivist. Inability to measure is sometimes just that — inability to measure — nothing more, nothing less. Einstein's finding says nothing meaningful with respect to the concept of absolute simultaneity. One is still perfectly free to choose, based on other considerations, whether or not the concept of true simultaneity is to be retained. And sensible understanding of experimental data, sensible derivation of the essential mathematics, is only possible when a universal absolute time and an underlying absolute simultaneity is retained.

Weinberg, Steven (1992) *Dreams of a Final Theory,* Pantheon Books New York, p 175.

46. James Gleick Julia Clark
> There will never be another Einstein. . . . Einstein's genius seemed nearly divine in its creative power: he imagined a certain universe and this universe was born.

Others would surely argue that striking the words 'seemed nearly' and replacing them with 'was', more in keeping with Einstein's achievement.

Gleick, James (1992) *Genius: The Life and Science of Richard Feynman*, Pantheon Books, New York, p 43.

47. Banesh Hoffmann — Julia Clark

> Practically all of the basic mathematical formulas of Einstein's 1905 paper on relativity are to be found in the 1904 paper of Lorentz and the two papers of Poincare, both of which latter warrant the date 1905 even though the major one did not appear until early 1906. The presence of often-identical formulas was almost inevitable, since relativity is intimately linked mathematically to Maxwell's equations and the mathematics of wave propagation. Indeed the mathematical transformation that is fundamental in relativity — a formula to which Poincare in 1905 gave the name *Lorentz transformation* — had already been found by the Irish-born physicist Joseph Larmor in 1898 on the basis of Maxwell's equations; and an almost identical transformation had been found by the German physicist Woldemar Voigt in a study of wave motion as early as 1887, the year of the Michelson-Morley experiment.
>
> These things, unfortunately, need to be said because the mathematical similarities have misled some people into the belief that Einstein's contribution was marginal, which it certainly was not. Yet in fairness we must add that among the writings of Poincare one finds so many of the relevant ideas that, with hindsight, one is surprised that he failed to take the crucial step that would have given him the theory of relativity, so close did he come to it.

Einstein's contribution to current theory, his legacy, was the corruption of science that remains so destructive of understanding to this day — his rejection of Lorentz's underlying asymmetry as 'intolerable' on positivistic grounds.

Hoffmann, Banesh (1972) Albert Einstein: Creator and Rebel, The Viking Press, New York, p 68.

48. Albert Einstein   Steven Mitchell
> Since the special theory of relativity revealed the physical equivalence of all inertial systems [absolute symmetry at all levels] it proved the untenability of the hypothesis of an ether at rest. It was therefore necessary to renounce the idea that the electromagnetic field is to be regarded as a state of a material carrier. The field thus becomes an irreducible element of physical description . . .

So an electromagnetic field is an 'irreducible element,' analogous to a vibration within a medium, but without the medium. An inconceivable, yet unavoidable, acknowledged requirement of SRT's perfect symmetry. Trouble is, it just ain't so.

Einstein, Albert, (1954) "Relativity and the Problem of Space," R.W. Lawson Trans. *Relativity, The Special and the General Theory*, Methuen and Company, Ltd., London

49. David Lindley   Steven Mitchell
> But the ether was by now a strikingly odd contrivance. . . . the interaction of matter with the ether had to have the peculiar property of altering the lengths of measuring sticks in such a way that any experiment designed to detect motion through the ether would be defeated and give a result indistinguishable from what would occur if the observer were truly at rest in the ether. The last hurrah of the mechanical ether models was thus a wholly unsatisfactory arrangement: the urge to preserve the stationary mechanical ether was so strong that an unexplained interaction between ether and matter was dreamed up with the sole purpose of thwarting any experimental attempt to detect it.

And he continues:

> . . . mechanical models of the ether were bound to

> fail, unless one was willing to introduce some arbitrary extra ingredient such as the Fitzgerald contraction.

Overall, a rather amusing supposed refutation of Lorentz's ether.

Lindley, David (1993) *The End of Physics*, BasicBooks, New York, p 45.

50. P. A. M. Dirac   Steven Mitchell
> Lorentz succeeded in getting correctly all the basic equations needed to establish the relativity of space and time.

So Dirac assured us that Lorentz developed "all the basic equations needed" within the confines of an ether theory, a theory that required the Fitzgerald contraction.

Gleick, James (1992) *Genius: The Life and Science of Richard Feynman*, Pantheon Books, New York, p 72.

51. Harold Fritzsch   Steven Mitchell
> We know that protons are extended objects, which we can view as little spheres . . . Now let's make a proton moving at almost the speed of light collide with another proton, or with a nucleus. What happens in this collision is quite complex, and I won't go into it here. But we do know that specific details of this process will be different if the impinging proton looks like a sphere or a disk.
>
> Some experiments of this kind were carried out at CERN. The results were unequivocal. Protons do behave like disks, and the faster they move, the flatter they get, just as [Einstein] predicted.

The Fitzgerald contraction (not, as Fritzsch implies, the 'Einstein contraction') is real (whatever that may mean) regardless of whether one goes with Einstein or holds to Lorentz.

Fritzsch, Harold (1994) *An Equation that Changed the World*, Karin Heusch, Trans., The University of Chicago Press, Chicago, p 151.

52. Eddington   Steven Mitchell
    When a rod is started from rest into uniform motion, nothing whatever happens to the rod.

But Fritzsch has assured us (note 51) that contraction is *real*. (Whatever that means.)

Eddington, Arthur S. (1918) *Report on the Relativity Theory of Gravitation*, Fleetway Press, London, p 8.

53. H. A. Lorentz   Steven Mitchell
    . . . there can be no question about the reality of this change of length.

The apparent disagreement between Eddington and Fritzsch seems to favor the *real* contraction of the Lorentz interpretation, but the SRT advocate can comfortably fall back on Hawking's observation that 'a physical theory is just a mathematical model and that it is meaningless to ask whether it corresponds to reality.' So what does 'real' mean? A physical reality (what really happens to the rod) is a meaningless concept since all we have is the measured values within Einstein's Universe. If you need a physical contraction to explain some result, you've got it. On the other hand, if only 'apparent,' works better with respect to other data, no problem — modern theory can accommodate.

Lorentz, H. A. (1921) *Nature* 106, Feb. 17, 1921, pp 793-795.

54. Paul Davies and John Gribbin   Steven Mitchell
    the reality exposed by modern physics is fundamentally alien to the human mind, and defies all power of direct visualization.

Ipso facto, any intuitively understandable explanation, any visualizable physical reality, must be rejected as in irreconcilable conflict with modern physics.

Davies, Paul and John Gribbin (1992) *The Matter Myth*, Simon & Schuster, New York, p 110.

55. Herbert Ives — Steven Mitchell

> The frequent assertion that the Michelson-Morley experiment abolished the ether is a piece of faulty logic. When Maxwell predicted a positive result from the experiment he did so on the basis of *two* assumptions; the first, that the light waves were transmitted through a medium, the second, which was not realized until pointed out by Fitzgerald, that the measuring instruments would not be affected by motion. The null result of the experiment proved *some* assumption made in predicting a positive result to be wrong. The experimental demonstration of the variation of measuring instruments with motion, in exactly the way to produce a null result, shows that it was the second assumption alone that was wrong; leaving the evidence for a transmitting medium, as derived from aberrational and rotational phenomena, as strong, if not stronger, than ever.

As Ives pointed out, the Michelson-Morley experiment revealed some unexpected truths, but left the evidence in favor of an ether as strong as ever. And, if one prefers a universally convergent sensible physical reality, then the modern understanding of the nature of matter tells us that a real physical medium is required.

Ives, Herbert E. (1948) "The Measurement of the Velocity of Light by Signals Sent in One Direction," *Journal of the Optical Society of America*, Vol. 38, Num. 10, October, 1948.

56. Albert Einstein　　　　　　　　Steven Mitchell
> Concerning the experiment of Michelson and Morley, H. A. Lorentz showed that the result at least does not contradict the theory of an ether at rest.

Just the facts. Lorentz gave us good theory, in no way contradicted by experiment, but fatally offensive to positivistic sensibilities.

Einstein, Albert, (1954) "Relativity and the Problem of Space," R.W. Lawson Trans. *Relativity, The Special and the General Theory*, Methuen and Company, Ltd., London

57. Albert Einstein　　　　　　　　Steven Mitchell
> Since the special theory of relativity revealed the physical equivalence of all inertial systems, it proved the untenability of the hypothesis of an ether at rest.

Einstein, far from using the Michelson-Morley experiment to justify the rejection of the Lorentz ether, used his own rejection of Lorentz's apparent symmetry in favor of absolute symmetry, his physical equivalence of all inertial systems, to abolish it.

Einstein, Albert, (1954) "Relativity and the Problem of Space," R.W. Lawson Trans. *Relativity, The Special and the General Theory*, Methuen and Company, Ltd., London

58. Steven Weinberg　　　　　　　　Steven Mitchell
> All these particles are bundles of the energy . . . of various sorts of fields. A field like an electric or magnetic field is a sort of stress in space, something like the various sorts of stress that are possible within a solid body, but a field is a stress in space itself.

One need only substitute 'ether' for Weinberg's 'space' and then explore the implications, to appreciate that the

Michelson-Morley results, far from disproving an ether, gave us the first experimental requirement for the current understanding of the nature of matter. The modern understanding, when coupled with MLET's worldview, tells us why matter must contract with increased velocity relative to the ether.

Weinberg, Steven (1992) *Dreams of a Final Theory,* Pantheon Books New York, p 25

59. D. H. Lawrence   Steven Mitchell

> I like relativity and quantum theories because I don't understand them and they make me feel as if space shifted about like a swan that can't settle, refusing to sit still and be measured; and as if the atom were an impulsive thing always changing its mind.

One finds this inclination to embrace the occult prevalent within the modern physics community. Some of this may be due to an attitude of "If you can't hide it, flaunt it," but many, with Lawrence, clearly cherish this aspect of Einstein's Universe.

Lederman, Leon with Dick Teresi (1993) *The God Particle,* Houghton Mifflin Company, New York.

60. Herbert Ives   Steven Mitchell

Ives argued that:

> The equality of the mass equivalent of radiation to the mass lost by a radiating body is derivable from Poincare's momentum of radiation (1900) and his principle of relativity (1904). The reasoning in Einstein's 1905 derivation, questioned by Planck, is defective. He did not derive the mass-energy relation.

And, in the same article, referring to a 1907 study by Planck, Ives wrote:

> This derivation [Planck's] of the relation . . . is historically the first authentic derivation of the relation.

Whether credit of origination is given to Einstein or others, it is generally acknowledged that Einstein first suggested it in the general form popular today. However, the fact that Einstein never cited it as one of his significant achievements indicates tacit recognition of the fact that, contrary to popular belief, he neither advanced nor hindered the general recognition of this matter-energy relationship to any significant degree.

Ives, Herbert E. (1952) "Derivation of the Mass-Energy Relation," *Journal of the Optical Society of America*, Vol. 42, Num. 8, August, 1952.

61. Lorentz, H. A. Steven Mitchell

> . . . this leads us to the idea that an atom is in the last resort some sort of local modification of the omnipresent ether . . .

and

> . . . we can therefore never set an electron in motion without simultaneously imparting energy to the ether. . . . in other words, if we determine the mass in the usual way from the phenomena, we get the true mass increased by an amount which we can call the apparent, or electromagnetic, mass. The two together form the effective mass which determines the phenomena.

So in 1902 we find Lorentz both anticipating the modern understanding of matter and clearly recognizing the increase of effective mass with increased velocity.

Lorentz, H. A. (1967) "The theory of electrons and the propagation of light," *Nobel Lectures: Physics: 1901-1921*, Elsevier Publishing Company, Amsterdam, pp 14-29

62. Steven Weinberg   Steven Mitchell

> All these particles are bundles of the energy . . . of various sorts of fields. A field like an electric or magnetic field is a sort of stress in space, something like the various sorts of stress that are possible within a solid body, but a field is a stress in space itself.

Weinberg's 'stress in space,' Lorentz's 'local modification of the ether.'

Weinberg, Steven (1992) *Dreams of a Final Theory*, Pantheon Books New York, p 25

63. P. A. M. Dirac   Steven Mitchell

> It is more important to have beauty in one's equations than to have them fit experiment.

One physicist/mathematician's view of the importance of mathematical beauty.

Davies, Paul (1992) *The Mind of God*, Simon & Schuster, New York, p 176.

64. Elie Zahar   Steven Mitchell

> Einstein differs from Lorentz in that he regards the 'effective' variables . . . as the real ones and totally abolishes the Galilean transformation. The [Lorentz's] Theory of Corresponding states is 'observationally equivalent' to Special Relativity [SRT] because experimental results involve only measured, that is, 'effective' quantities. Since the latter satisfy Maxwell's equations, we are unable, whether we adopt Lorentz's or Einstein's theory, to decide on empirical grounds whether our frame of reference is in motion or at rest in the 'ether'.

And

> As I have already shown, Lorentz's theory is observationally equivalent to the SRT; Einstein's transformed coordinates can be interpreted as the

measured coordinates in Lorentz's moving frame. In the latter the 'real' coordinates are still the Galilean ones.

Zahar pointed out in the same speech that an important heuristic rule required by Einstein's interpretation is that one must:
> replace any theory which does not explain symmetrical observational situations as the manifestations of deeper symmetries — whether nor not descriptions of all known facts can be deduced from the theory.

Zahar, Elie, "Why did Einstein's Programme supersede Lorentz's?" *Method and appraisal in the physical sciences: The critical background to modern science, 1800-1905*, Colin Howson Ed., Cambridge University Press, Cambridge, pp 211-275

65. Reza Mansouri and Roman Sexl   Steven Mitchell

They have written three papers in which they constructed
> an ether theory . . . that maintains absolute simultaneity and is kinematically equivalent to special relativity.

They tell us that the basis of an alternative to SRT is rigorous and real.
> All experiments can be explained either on the basis of special relativity or by an ether theory . . . This demonstrates . . . the impossibility of an "experimentum crucis" deciding between ether theories and the special theory of relativity.

So direct experiment can't tell the difference. Nevertheless, there is a real difference.
> . . . the symmetry group contained in Einstein's theory restricts the possible forms of electrodynamic and other interaction so strongly that on the basis of Lorentz invariance alone an

> interaction that is known in one system of reference can be rewritten in any other. . . . In other words, the symmetry group contained in relativity makes many predictions possible, which have to be derived with the help of additional assumptions in ether theories.

And the disallowed additional assumptions give us our sensible, physical reality. Because of the way the underlying asymmetry is reflected in the measured symmetry, pragmatically, with an ether theory, we are perfectly free to choose to consider any arbitrary inertial frame to be the ether frame. But, as Mansouri and Sexl point out:

> By singling out arbitrarily one system . . . to be the ether system one destroys the equivalence of all inertial systems . . .

And that, Einstein could not tolerate. This intolerable asymmetry is the asymmetry that gives us a sensible, physical reality and good explanation.

Mansouri , Reza and Roman U. Sexl (1977) "A Test Theory of Special Relativity: I. Simultaneity and Clock Synchronization" *General Relativity and Gravitation* (USA) Vol. 8 No. 7, pp 497-513.

Mansouri , Reza and Roman U. Sexl (1977) "A Test Theory of Special Relativity: II. First Order Tests" *General Relativity and Gravitation* (USA) Vol. 8 No. 7, pp 515-524.

Mansouri , Reza and Roman U. Sexl (1977) "A Test Theory of Special Relativity: III. Second-Order Tests" *General Relativity and Gravitation* (USA) Vol. 8 No. 10, pp 809-814.

66. Albert Einstein   Steven Mitchell

> For the theoretician such an asymmetry in the theoretical structure, with no corresponding asymmetry in the system of experience, is

intolerable.

Modern physics theory has followed his lead.

Einstein, Albert (1983) "Ether and The Theory of Relativity," (An address delivered on May 5th, 1920, in the University of Leyden) *Sidelights on Relativity*, Dover Publications, Inc, New York, p 12.

67. Stanislaw Ulam   Steven Mitchell
    I told Fermi how in my last year of high school I was reading popular accounts of the work of Heisenberg, Schrodinger, and De Broglie on the new quantum theory. I learned that the solution of the Schrodinger equation gives levels of hydrogen atoms with a precision of six decimals. I wondered how such an artificially abstracted equation could work to better than one part in a million. A partial differential equation pulled out of thin air, it seemed to me, despite the appearances of derivation by analogies. I was relating this to Fermi, and at once he replied: "It [the Schrodinger equation] has no business being that good, you know, Stan.

Why does it work so well?

Ulam, Stanislaw (1993) "Los Almos," *The World Treasury of Physics, Astronomy and Mathematics*, Timothy Ferris Ed., Little, Brown and Company, Boston, pp 724-725. (Reprinted from Ulam's (1976) *Adventures of a Mathematician*, Charles Schribner's Sons, New York)

68. Albert Einstein   Steven Mitchell
    It is often, perhaps even always, possible to adhere to a general theoretical foundation by securing the adaptation of the theory to the facts by means of artificial additional assumptions.

Securing adaptation of new facts to SRT's perfect symmetry, to the "general theoretical foundation" is such a challenging task that physicists have little time or inclination to re-examine the foundation — especially, given that they know that that foundation is absolute *fact*.

Einstein, Albert (1993) "Autobiographic Notes," *The World Treasury of Physics, Astronomy and Mathematics*, Timothy Ferris Ed., Little, Brown and Company, Boston, p 585.

69. Steven Weinberg Steven Mitchell
    There is nothing in any single disagreement between theory and experiment that stands up and waves a flag and says, "I am an important anomaly."

The flags are up, and the LIGO failure to detect gravitational waves may provide the strong breeze necessary to set them waving vigorously.

Weinberg, Steven (1992) *Dreams of a Final Theory*, Pantheon Books New York, p 94.

70. P. A. M. Dirac Steven Mitchell
    One can thus make quantum electrodynamics into a sensible mathematical theory, but only at the expense of spoiling its relativistic invariance.

Dirac personally found the infinities of quantum electrodynamics troublesome, and, although as indicated, he found a reasonable solution, he just couldn't bring himself to give up SRT's invariance demands and so concluded that something must be wrong with quantum theory. Since the problem could never be traced to quantum theory, the current consensus is that (since SRT itself is 'fact') and since one must, at all costs, secure 'adaptation of the theory to the facts,' one must give up any prejudice in favor of strict mathematical integrity. Dubious mathematics, yes, but we still have a theory that predicts the results of events that can be measured. So much for mathematics as a sure guide — even mathematical integrity must give way when required to do so by SRT's prevailing absolute.

Dirac, P. A. M. (1973) *Directions in Physics*, John Wiley & Sons, New York, p 36.

71. James Clerk Maxwell Steven Mitchell
    Hence all these theories lead to the *conception of a medium in which the propagation takes place*, and if we admit this medium as an hypothesis, I think it ought to occupy a prominent place in our investigations, and that we ought to endeavor to construct a mental representation of all the details of its action, *and this has been my constant aim in this treatise* (emphasis added).

Einstein went so far as to suggest that it was Maxwell who first relieved us of the necessity of an ether by reducing electromagnetic radiation to interaction of fields, fields mathematically independent of the necessity of any medium. It seems clear that Maxwell would not have welcomed Einstein's rejection of the validity of his stated intent.

Maxwell, James Clerk (1954) *A Treatise on Electricity and Magnetism*, Third Edition, Dover Publications, Inc. Vol. 2, p 493.

72. Richard Feynman Julia Clark
    You can recognize truth by its beauty and simplicity. When you get it right, it is obvious that it is right — at least if you have any experience — because usually what happens is that more comes out than what goes in . . . The inexperienced, the crackpots, and people like that, make guesses that are simple, but you can immediately see that they are wrong, so that does not count. Others, the inexperienced students, make guesses that are very

> complicated, and it sort of looks all right, but I know it's not true because the truth always turns out to be simpler than you thought.

'More comes out than goes in.' Start with a sensible underlying asymmetry and get back universal convergence and a sensible physical reality. I like it.

Kaku, Michio (1994) *Hyperspace*, Oxford University Press, New York, p 130

73. Ron Hatch   Steven Mitchell
> It is also apparent from the model that the correct, but unusual, relationship between spin and magnetic dipole moment will result. The three pairs of spinning faces combine as a vector to give the total spin value, but half of the spin is canceled at the common boundaries of the faces. The angular relationships are such that the spin of the total particle is slightly less than the spin of one pair of opposite faces. The spin of a pair of faces is the inverse of the square root of three, while the spin of the entire structure is one-half (in units of Planck's constant over 2 pi). The magnetic dipole of each pair of opposite faces adds as a vector without any canceling effect. Thus the approximate g-factor (very close to two) of the electron is obtained directly. The magnetic dipole moment is twice as big as it would be if the structure were spinning as one combined whole.

Discussing the MLET model of the electron.

Hatch, Ronald R. (1992) *Escape From Einstein*, Kneat Kompany, Wilmington, CA, p 141.

74. Ron Hatch   Steven Mitchell
> But a strange thing happens to this particle on its way to reality. The phase of the negative north pole still moves outward at the speed of light, and the

phase of the positive south pole still moves inward at the speed of light Thus the energy of the particle will be minimized if it moves in the downward direction (opposite to its spin vector) at the speed of light or at least very close to the speed of light.

Discussing the possible specific structure of a neutrino.

Hatch, Ronald R. (1992) *Escape From Einstein*, Kneat Kompany, Wilmington, CA, p 142.

75. Ron Hatch   Steven Mitchell
The most significant improvement in understanding diverse quantum phenomena is, I believe, the movement from "particle" to "standing-wave" pictures of physics. This provides for non-locality and probability wave collapse that remains profoundly puzzling from a particle point of view.

One of the significant benefits of the recognition of an underlying asymmetry.

Hatch, Ronald R. (1992) *Escape From Einstein*, Kneat Kompany, Wilmington, CA, p 150.

76. Michio Kaku   Steven Mitchell
In fact, the amount of energy flooding the vacuum is 10^{100} times larger than the experimentally observed amount. In all of physics, this discrepancy of 10^{100} is unquestionably the largest.

With MLET's gravity field, a density gradient field, we require the measured results.

Kaku, Michio (1994) *Hyperspace*, Oxford University Press, New York, p 267

77. Elie Zahar   David Rhodes
Lorentz was justified in asserting that:
" . . . the chief difference [is] that Einstein *simply postulates* what we have *deduced* . . . from the

fundamentals of the electromagnetic field."

> Lorentz explained Michelson's result in a non *ad hoc* way; he was first to discover the transformation laws for the electromagnetic field; he described the way in which the inertia of the electron depends both on its energy and on its velocity; and he explained the invariance of c. Thus Lorentz's continued adherence to his own programme after 1905 was completely rational.

> Lorentz's theory was eminently intelligible whereas Einstein's involved a major revision of our most basic notions of space and time.

These were key points in convincing me that SRT is inferior to MLET. I expect that, as MLET predicts, the LIGO observatories will detect nothing.

Zahar, Elie, "Why did Einstein's Programme supersede Lorentz's?" *Method and appraisal in the physical sciences: The critical background to modern science, 1800-1905*, Colin Howson Ed., Cambridge University Press, Cambridge, pp 211-275

78. Stephen Hawking   David Rhodes

> There is a subspecies called philosophers of science who ought to be better equipped. But many of them are failed physicists who found it too hard to invent new theories and so took to writing about the philosophy of physics instead. They are still arguing about the scientific theories of the early years of this century, like relativity and quantum mechanics. They are not in touch with the present frontier of physics.

Settled beliefs are not open to question. It will be very interesting to see how the null LIGO results will be handled by the consensus community.

Hawking, Stephen (1993) *Black Holes and Baby Universes and Other Essays*, Bantam Books, New York, p 41.

ABOUT THE AUHTOR

Ed Hatch, near the end of a thirty-year career in the computer sciences, became aware of his brother Ron's Modified Lorentz Ether Theory (MLET) and found it too fascinating to put aside. After retiring in 1991, he has since spent most of his free time examining the implications of the theory. This involved reading extensively and included development of some computer simulations of some fundamental phenomena based on the MLET model.

Since MLET predicts some very explicit and generally unexpected results with regard to some planned experiments, Ed felt it important to make these predictions known before the experiments are fully operational. Ed wrote this book to create awareness of the MLET predictions and to introduce the basic theory.

The author of the theory is Ed's brother, Ron. Ron Hatch is the current chair (an elected office) of the Satellite Division of the Institute of Navigation. This is the organization that conducts the premier conference relating to GPS, a conference typically drawing over 2,000 people to the sessions and exhibits.

Ron has been working with navigation and communications using satellites since 1962, when, still in college, he worked for the U.S. Science Exhibit at the Seattle World's Fair demonstrating the Doppler effect on the signals received from the TRANSIT satellites of the Navy Navigation Satellite System. This system was developed by Johns Hopkins Applied Physics Laboratory, where Ron worked developing navigation algorithms immediately following college.

In late 1993, after many years of working for others, Ron began what proved to be a very successful private consultation practice that included such clients as NASA, FAA, Motorola, and Leica, the Swiss survey company. In 1995, he, along with four other consultants, started NavCom Technology, which has grown into a successful GPS and satellite communications company employing more than 50 people. Although full time at

NavCom since 1997, Ron has maintained some independent consultation work with others.

In 1994, Ron was awarded the Johannes Kepler award for 'Sustained and Significant Contributions to Satellite Navigation' from the Satellite Division of the Institute of Navigation - only the fourth recipient of this prestigious award. These 'sustained and significant' contributions were made over many years. He's received eight patents relating to GPS and satellite navigation, with several more in process. Among contributions not patented, is a technique Ron developed for removing much of the noise caused by electromagnetic reflections from the fundamental Global Positioning System measurements. This technique is now employed in virtually all GPS receivers and is referred to in FAA algorithm documents as the 'Hatch filter.'

Ron's understanding of the GPS system includes awareness of the effects of gravity and velocity on precision atomic clocks and other important relativity effects. His Modified Lorentz Ether Theory came from a driving preference for rigorous understanding, rather than from any personal desire to significantly impact theoretical physics. But in the course of that effort he became convinced that acceptable understanding of relativity effects could only come with a radical departure from consensus thinking. MLET, in its current form (including the prediction that LIGO will not detect gravitational waves) has far surpassed his original, rather modest goal.

www.ingramcontent.com/pod-product-compliance
Lightning Source LLC
Chambersburg PA
CBHW030917180526
45163CB00002B/366